International Library of and the New Medicine

Volume 70

Series editors
David N. Weisstub, University of Montreal Fac. Medicine Montreal, QC, Canada
Dennis R. Cooley, North Dakota State University, History, Philosophy, and Religious Studies, Fargo, ND, USA

The book series International Library of Ethics, Law and the New Medicine comprises volumes with an international and interdisciplinary focus. The aim of the Series is to publish books on foundational issues in (bio) ethics, law, international health care and medicine. The 28 volumes that have already appeared in this series address aspects of aging, mental health, AIDS, preventive medicine, bioethics and many other current topics. This Series was conceived against the background of increasing globalization and interdependency of the world's cultures and governments, with mutual influencing occurring throughout the world in all fields, most surely in health care and its delivery. By means of this Series we aim to contribute and cooperate to meet the challenge of our time: how to aim human technology to good human ends, how to deal with changed values in the areas of religion, society, culture and the self-definition of human persons, and how to formulate a new way of thinking, a new ethic. We welcome book proposals representing the broad interest of the interdisciplinary and international focus of the series. We especially welcome proposals that address aspects of 'new medicine', meaning advances in research and clinical health care, with an emphasis on those interventions and alterations that force us to re-examine foundational issues.

More information about this series at http://www.springer.com/series/6224

Pedro Serna • José-Antonio Seoane
Editors

Bioethical Decision Making and Argumentation

 Springer

Editors
Pedro Serna
Universidad Internacional de La
   Rioja (UNIR)
Logroño, Spain

Research Group Philosophy, Constitution
   and Rationality
Universidade da Coruña
A Coruña, Spain

José-Antonio Seoane
Research Group Philosophy, Constitution
   and Rationality
School of Law
Universidade da Coruña
A Coruña, Spain

ISSN 1567-8008   ISSN 2351-955X   (electronic)
International Library of Ethics, Law, and the New Medicine
ISBN 978-3-319-82819-0   ISBN 978-3-319-43419-3   (eBook)
DOI 10.1007/978-3-319-43419-3

© Springer International Publishing Switzerland 2016
Softcover reprint of the hardcover 1st edition 2016
This work is subject to copyright. All rights are reserved by the Publisher, whether the whole or part of the material is concerned, specifically the rights of translation, reprinting, reuse of illustrations, recitation, broadcasting, reproduction on microfilms or in any other physical way, and transmission or information storage and retrieval, electronic adaptation, computer software, or by similar or dissimilar methodology now known or hereafter developed.
The use of general descriptive names, registered names, trademarks, service marks, etc. in this publication does not imply, even in the absence of a specific statement, that such names are exempt from the relevant protective laws and regulations and therefore free for general use.
The publisher, the authors and the editors are safe to assume that the advice and information in this book are believed to be true and accurate at the date of publication. Neither the publisher nor the authors or the editors give a warranty, express or implied, with respect to the material contained herein or for any errors or omissions that may have been made.

Printed on acid-free paper

This Springer imprint is published by Springer Nature
The registered company is Springer International Publishing AG Switzerland

Foreword

The roots of *Bioethical Decision-Making and Argumentation* lie in a research project funded by the Spanish government[1] and carried out by a group of Philosophy of Law lecturers and researchers with a long-standing interest in bioethics and in the theory and practice of argumentation, especially in legal contexts. The goals of this project were threefold:

1. First, to provide a critical analysis of the prevailing paradigm in bioethical decision-making, namely, North American principlism, propounded by Beauchamp and Childress
2. Second, to provide a similar critical analysis of other proposals arising from the debate surrounding the above
3. Finally, to put forward the fundamental lines of an alternative model combining substantive principles and procedural guidelines, robust enough not only to overcome the shortcomings of principlism and its alternatives but also to stand up to the criticisms usually levelled at ethics, regardless of whether they focus on moral contents or merely on discursive procedures

The initial hypothesis is that the principlist model of decision-making is open to improvement and replacement by a model combining substantive principles and procedural guidelines, the former consisting of a set of basic rights that are part of the Western cultural heritage and the latter of criteria relating to prudent discursive practical reasoning. Such a model makes it possible to conserve the more valuable elements of the principlist model whilst avoiding some of the pitfalls highlighted by its critics.

The team of academics working on the project took as their starting point the close links between bioethics and law, not only in the historical or genetic sense,

---

[1] Research project *Principialismo y teorías de la argumentación en la toma de decisiones médicas* (DER2010-17357), led by Professor Pedro Serna and funded by the Spanish Ministry of Science and Innovation.

which is hardly open to discussion, but also in a much deeper sense. Indeed, in modern Western societies, the goods at stake in the doctor-patient relationship are protected by law and usually at the highest possible level, that of the constitution. This in turn permits us to suppose that the principles and forms of argumentation developed in the field of the law are able to throw light on the task of constructing a bioethical decision-making model, with particular reference to the clinical sphere, able to withstand the most pertinent criticisms made against principlism. Thus, the contemporary debate on principles and their application to individual cases, normative systems theory, balancing theory and its practice and, more broadly, argumentation theories together with the constitutional dogmatics of basic rights supplies useful elements with which to design a biomedical decision-making model containing both substantive elements (human rights) and procedural ones (discursive argumentation and prudential models).

***

Ever since it was first formulated in the United States in the late 1970s, principlism has been the paradigm for biomedical ethics, particularly in the clinical field. This, however, has not spared it from numerous critiques attacking its ambiguity and incompleteness and putting forward alternatives such as casuistry, virtue bioethics or the model based on so-called common morality. Moreover, a specific form of principlism has developed in Europe that proposes replacing the principles of non-maleficence, beneficence and justice with those of dignity, integrity and vulnerability, whilst retaining that of respect for autonomy. Some authors have proposed other modifications to principlism, such as the introduction of hierarchies governing the application of principles, whilst others go further, assuming many of the critiques as valid and proposing a more flexible model that replaces the four principles with a decision-making process based on balancing values.

One of the common characteristics of modern Western societies is that they are home to a variety of groups and individuals who hold different and partially incompatible moral ideas ("comprehensive doctrines", according to the expression coined by Rawls,[2] or "moral communities", in Engelhardt's words[3]). As is well known, the existence of this radical pluralism poses some of the biggest challenges faced by the theory of justice and, more widely, by the law today. Bioethical problems are no exception to this and indeed even constitute the field in which the conflicts deriving from this irreducible pluralism are most sharply expressed, and thus the one in which the models used to channel and resolve them must prove their effectiveness most robustly. Similarly, just as in the medieval world it was inconceivable to think of ethics without the idea of God and salvation, in the modern world, it is impossible to conceive of models of social ethics that fail to take this pluralism into account. For some authors, Rawls amongst them, this implies a political rather than metaphysical idea of justice and social ethics in general.

---

[2] *Political Liberalism*, Columbia University Press, New York, 1993.
[3] *The Foundations of Bioethics*, 2nd ed., Oxford University Press, New York, 1996.

Foreword

Furthermore, bioethics is specifically destined to be the source of the paradigm of ethics in modern society, precisely because it deals with problems arising from scientific and technological progress, coming at a time when the theocentric world-view is fast losing currency.[4]

A specific feature of modern bioethics is that generally speaking the problems it poses arise not from a traditional deontological perspective, i.e. that of a doctor or health professional's moral obligations, but from a relational standpoint, i.e. within the framework of the relationship between health professional and patient or user of health services, which are usually considered as public services in the social and democratic constitutional state that is generally accepted as the model in the societies we are referring to. This obliges us to consider whether bioethics is really, as some insist, part of morality or ethics in the traditional sense or, on the contrary, whether the problems it deals with are inevitably legal ones, since the goods at stake in the relation referred to above (life, physical and moral integrity, health, personal autonomy, etc.) are recognised and protected by law in the whole of the Western world.

It therefore comes as no surprise that proposals for the "juridification of bioethics" have been made as a consequence of the above.[5] Indeed, the origins of bioethics as a discipline cultivated by a professional and academic community are legal ones in the countries in which it arose: to cite only the cases of the United States and Canada, the legal system itself has encouraged, if not forced, this juridification in the form of a judicialization that should be credited as having been the starting point for bioethics considered as an academic discipline.[6] This, however, is not without its perverse effects or consequences; a case in point would be the way in which it has modified the doctor-patient relationship, traditionally characterised by features of debatable value, such as paternalism, but also by others whose value is beyond discussion, such as trust, and which are nowadays severely threatened, especially in Western societies.

Juridification, in the sense of judicialization, allows the great majority of bioethical conflicts to be resolved a posteriori but at the same time fails to provide social and healthcare professionals with a decision-making model enabling them to decide their course of action within a time frame that is at best limited and at worst almost non-existent. This may be the reason behind the tendency to exacerbate the principle of patient autonomy as the key governing factor in the clinical relationship, with the de facto consequence that it functions more as a defence mechanism or a means of offloading responsibility on the part of health professionals than as an expression of respect for a patient's autonomy or a manner of exercising informed consent. For

---

[4] Cf. Drane, "What is Bioethics? A History", in Lolas and Agar (eds.), *Interfaces Between Bioethics and the Empirical Social Sciences*, Regional Program on Bioethics OPS/OMS Publication Series, Santiago de Chile, 2002, downloadable from http://www.paho.org/English/BIO/interfaces.pdf.

[5] Atienza, "Juridificar la bioética. Una propuesta metodológica", *Claves de razón práctica* 61 (1996).

[6] On the influence of the law on the shaping of bioethics, see Annas, *Standard of Care. The Law of American Bioethics*, New York, Oxford University Press, 1993.

this reason, the most appropriate form of juridification would appear to move in the direction of the contribution that the modern theory of law and legal argumentation can make as a reasoning tool in the decision-making process.

There are those who consider bioethics to be based on substantive principles, normally originating from comprehensive views of the world, mankind and human existence that are very often linked, either implicitly or explicitly, to religious outlooks on life. However, the prevailing paradigm in modern-day biomedical ethics is the one put forward by T.L. Beauchamp and J.F. Childress (henceforth Beauchamp and Childress) in *Principles of Biomedical Ethics* (1979), now considered a classic work in the field and widely endorsed in both academic and professional circles. Nevertheless, and in spite of its widespread acceptance, the 1990s saw the appearance of intense debate and criticism of this model, causing its authors to modify their stance in subsequent editions, of which there are now seven, the most recent of which was published in 2013. Curiously enough, this debate has until now been circumscribed to the North American social and cultural stage, even though European authors are involved in it.

Beauchamp and Childress' proposal (henceforth principlism) is based on a set of ideas which clearly display the influence of John Rawls, an author whose work they cite profusely; they include the irreducible nature of the substantive moral concepts that co-exist in our societies and the numerous ethical and philosophical theories that currently abound, rendering any kind of bioethics based on material principles unworkable and making it necessary to base proposals for bioethical decision-making models on intermediate-level principles.[7] Although the content of these intermediate principles is substantive, they show a noticeable tendency to operate in a formal and procedural way. According to Beauchamp and Childress, the principles in question are respect for autonomy, non-maleficence, beneficence and justice, and their model deals fundamentally with the decision-making process, leaving little room for what would have constituted the basic elements of a more traditional view of medical deontology, namely, the ethos of doctor-patient relations, on the one hand, and the virtues of health professionals, on the other. Although Beauchamp and Childress deal with both of these matters, they do so only marginally.

The principlist proposal has been widely scrutinised in the bioethical literature, being criticised by some authors whilst others defend it or suggest amendments. The critics who propose alternative models to principlism can be divided into two camps: those who approach the subject from the point of view of casuistry and those who do so from the standpoint of so-called common morality. Both schools of thought, however, generally coincide in pointing out the lack of functionality of the principlist model, attributing it to the fact that the principles listed above are by themselves insufficient to create a normative system endowed with unity, consistency and completeness (or fullness).

---

[7] In the sense put forward by Wildes, "Principles, Rules, Duties, and Babel: Bioethics in the Face of Postmodernity", *Journal of Medicine and Philosophy* 25/3 (2000) and subsequently accepted by Beauchamp and Childress in the 5th edition of their classic work (2001), p. 407.

Casuistry[8] has its roots in both the theories of moral casuistry and the Anglo-North American legal tradition, based on precedent. True to its origins, casuistry makes no claim to construct a system and instead proposes a problem-solving method that is not without issues of its own, including those deriving from the constantly shifting scenario and the constant novelty that is a characteristic of ethical problems, linked to the ongoing progress of life sciences in general and biotechnology in particular.

For its part, so-called common morality[9] holds that intermediate principles are unable to fulfil the objectives for which they were created, namely, to take the place of substantive moral theories and act as guidelines for deciding on the morally correct course of action. The authors who take this view stress principlism's lack of unity and consistency and highlight its conceptual and epistemic gaps, precisely with regard to what exactly constitutes a principle and how to work with them. With this as their starting point and drawing the distinction between morality and moral theory, they propose a model based on the former in which public or commonly accepted morality becomes the source for bioethical criteria, since it is recognisable and acceptable to any serious moral agent.

Beauchamp and Childress, together with some of their followers, have based their response to their critics on a strategy of eclectically integrating a selection of the proposals put forward as critiques of their theory. For example, in order to redress the lack of functionality and consistency, they have seen fit to add two further elements to the principlist model: specification and balancing. Specification has mainly been dealt with by Richardson,[10] whilst the authors who have developed the concept of balancing include, amongst others, DeMarco and Ford.[11] Beauchamp, meanwhile, has answered his critics directly in a number of journals,[12] even though some of the suggestions relating to balancing and specification were added to the principlist model in the fifth edition of *Principles of Biomedical Ethics* (2001). Curiously enough, it is the very lack of precision and consistency denounced by

---

[8] Cf., for example, Jonsen and Toulmin, *The Abuse of Casuistry*, Berkeley, University of California Press, 1988; Arras, "Getting Down to Cases: The Revival of Casuistry in Bioethics", *Journal of Medicine and Philosophy* 16 (1991); Wildes, "The Priesthood of Bioethics and the Return of Casuistry", *Journal of Medicine and Philosophy* 18/1 (1991); Strong, "Specified Principlism: What Is It, and Does It Really Resolve Cases Better than Casuistry?", *Journal of Medicine and Philosophy* 25/3 (2000); and Jonsen, "Strong on Specification", *Journal of Medicine and Philosophy* 25/3 (2000).

[9] Cf., for example, Clouser and Gert, "A Critique of Principlism", *Journal of Medicine and Philosophy* 15 (1990); Green, Gert and Clouser, "The Method of Public Morality vs. the Method of Principlism", *Journal of Medicine and Philosophy* 18 (1993).

[10] Cf., for example, Richardson, "Specifying, Balancing, and Interpreting Bioethical Principles", *Journal of Medicine and Philosophy* 25/3 (2000).

[11] "Balancing in Ethical Deliberation: Superior to Specification and Casuistry", *Journal of Medicine and Philosophy* 31 (2006).

[12] Cf. initially, Beauchamp, "Reply to Strong on Principlism and Casuistry", *Journal of Medicine and Philosophy* 25/3 (2000); Beauchamp, "Methods and Principles in Biomedical Ethics", *Journal of Medical Ethics* 29 (2003); and the 5th (2001), the 6th (2009) and the 7th (2013) editions of Beauchamp and Childress, *Principles of Biomedical Ethics*, Oxford University Press, New York.

their critics that makes their strategy possible, and in this regard, it can be considered far from satisfactory. It therefore makes sense to reiterate a number of these criticisms, as is done in some of the chapters in this book.[13]

Finally, proposals have also been made to steer the biomedical decision-making issue in the direction of professional deontology and/or virtue ethics[14] or to adopt an approach based on Gadamerian hermeneutics.[15]

Europe has also been strongly influenced by North American principlism but not without significant criticism.[16] Furthermore, it has been home to the development of a different principlist model involving four principles (autonomy, dignity, integrity and vulnerability) of which only the first coincides with those listed in North American principlism.[17] The mere enumeration of these principles reveals the different moral sensitivity to which they owe their origin (particularly in the case of dignity and vulnerability), although this only serves to confirm, rather than eradicate, the epistemological, systematic and conceptual difficulties highlighted by the critics of North American principlism.

---

[13] An overview of this debate can be found in Smith Iltis, "Bioethics as Methodological Case Resolution: Specification, Specified Principlism and Casuistry", *Journal of Medicine and Philosophy* 25/3 (2000) and Davies, "The Principlism Debate: a Critical Overview", *Journal of Medicine and Philosophy* 20 (1995). Amongst authors writing in Spanish, cf. Ferrer and Álvarez Pérez, *Para fundamentar la Bioética: teorías y paradigmas teóricos en la Bioética contemporánea*, Bilbao, Desclée de Brouwer, 2003; Requena Meana, "Sobre la aplicabilidad del principialismo norteamericano", *Cuadernos de Bioética* 65 (2008); and Campos Serena, "Bioética principialista. El papel de la tradición norteamericana", downloadable at http://cfj.filosofia.net/2008 (last accessed 16 January 2016).

[14] Cf., for example, Pellegrino and Thomasma, *The Virtues in Medical Practice*, New York-Oxford, Oxford University Press, 1993.

[15] Cf. Lingiardi and Grieco, "Hermeneutics and The Philosophy of Medicine: Hans-Georg Gadamer's Platonic Metaphor", *Theoretical Medicine and Bioethics* 20 (1999); Svenaeus, "Hermeneutics of Clinical Practice: The Question of Textuality", *Theoretical Medicine and Bioethics* 21 (2000); and, by the same author, "Hermeneutics of Medicine in the Wake of Gadamer: the Issue of Phronesis", *Theoretical Medicine* 24 (2003), together with their accompanying references.

[16] In the Spanish context, the most noteworthy critical approach is that adopted by Gracia Guillén, who is undoubtedly the most influential author writing on bioethics in Spanish. From a standpoint that he himself defines as Aristotelian, Gracia at first accepted the postulates of principlism, although he established different levels between the principles concerned (cf. Gracia, *Fundamentos de Bioética*, Madrid, Eudema, 1989; Gracia, *Procedimientos de decisión en ética clínica*, Madrid, Eudema, 1991), but subsequently considered it too narrow, proposing that balancing should be carried out with regard to values rather than principles (cf. Gracia, "La deliberación moral: el método de la ética clínica", *Medicina Clínica* 117 (2001); Gracia, *Como arqueros al blanco. Estudios de Bioética*, San Sebastián, Triacastela, 2004).

[17] The proposal was published in 2002, as the outcome of a European BIOMED II research project (1995–1998) involving 22 professionals from a number of European Union member states. Cf. Rendtorff, "Basic Ethical Principles in European Bioethics and Biolaw: Autonomy, Dignity, Integrity and Vulnerability – Towards a Foundation of Bioethics and Biolaw", *Medicine, Health Care and Philosophy* 5 (2002); "Update of European Bioethics: Basic Ethical Principles in European Bioethics and Biolaw", *Bioethics UPdate* 2 (2015).

Foreword   xi

As we have seen, the debate is far from over on both sides of the Atlantic. Furthermore, at times insufficient attention is paid to the fact that the answers to a number of the problems, and even the problems themselves, have to be formulated in the light of the significant contextual differences between Europe and the United States in the field of healthcare. These differences include but are not necessarily limited to:

(a) The lack of a universal social healthcare model in the United States, in contrast to the universal social security model that prevails in Europe as an expression of the social and democratic constitutional state
(b) The prevalence of an individualistic mentality compared to the family- and community-based mentality that still prevails in Europe, particularly in Mediterranean countries
(c) A heightened liberalism, which leads to the doctor-patient relationship being seen as a private legal relation between individuals, with consequences that are considerably aggravated by the civil liability system characteristic of North American law, whilst in Europe healthcare enjoys the status of a public service, so that on the whole clinical and caregiving relations fit within an administration-constituent relational framework and the civil liability system lacks the punitive dimension to be found on the other side of the Atlantic
(d) The differences in mentality regarding the technique of legal and practical reasoning between the United States, where the common law method applies, and the majority of European countries, where practical rationality seeks to base decisions on more or less universal criteria (general norms of a legal, ethical or other nature)

***

Bioethical Decision-Making and Argumentation owes its structure to a combination of the above-mentioned considerations and the order in which its chapters appear. The first three of these, which are general and expository in nature, provide a framework for debate and a context for the subsequent chapters, which introduce principlism (Chap. 1) and two of the main alternative biomedical decision-making methodologies: a deliberative and value-based approach (Chap. 2) and a human rights-based approach (Chap. 3). The next two chapters offer a critique of the prevailing principlist model from a dual perspective: its lack of sufficient grounding and functionality (Chap. 4) and its systematic shortcomings (Chap. 5). Chapters 6 and 7 highlight the importance of characterising both action and method (in this case deliberative) in biomedical decision-making processes, stressing the relevance of practical reasoning. The final two chapters of the book deal with the legal and institutional aspects of the debate: Chap. 8 appeals to the principle of proportionality or reasonability, from which standpoint it interprets the principle of double effect, whilst Chap. 9 considers the way in which the previous contributions act on decision-making processes taking place within the institutional framework of bioethics committees.

The book opens with an updated presentation of principlism by its leading representative, Tom L. Beauchamp, who in Chap. 1 describes the salient feature of his four-principle model, enriched by having incorporated the main criticisms levelled at it from the standpoints of virtue ethics and, above all, common morality, together with its use of specification to resolve concrete ethical problems in biomedical practice and research. In Chap. 2, Diego Gracia puts forward an alternative approach to that of the previous model, based from a substantive standpoint on the triad of facts, values (as opposed to principles) and duties and from a procedural one on deliberation, with its roots in Aristotelian philosophy. A view that comes closer to European principlism is presented in Chap. 3, where Roberto Andorno approaches bioethical and biotechnological problems from the human rights angle, describing its principal normative instruments at international level, the reasons for adopting such a perspective and, finally, its principal shortcomings, which are most visible in the biotechnology sphere.

In Chap. 4, Carolina Pereira highlights the difficulties principlism faces in justifying the moral norms underpinning bioethics, seen as a set of universally shared moral beliefs, and in providing rational guidelines for action in the biomedical field: in her view, the principlist model is handicapped by a certain degree of intuitionism and its abandonment of rational justification. Chapter 5, on the other hand, offers a critique of principlism in which Óscar Vergara approaches the issue from the perspective of normative systems, which apply to both ethics and the law. His analysis, of a kind not commonly encountered, reveals certain systematic shortcomings (e.g. lack of completeness, inconsistency and partial indeterminacy) that not only make it impossible for it to be taken as a comprehensive biomedical decision-making model but also render it unworkable in certain cases.

Pilar Zambrano devotes Chap. 6 to an aspect of the problem that instead of principles takes as its starting point the classification of actions, an issue that bioethical theories tend to ignore. If the purpose of ethics is to guide actions and determine their correctness, it is therefore essential to individualise and classify them in order to establish a correspondence between actions and principles or values. The author develops this view from the theoretical standpoint and provides a working example in the form of life-saving care, closely following Anscombe's thinking, with its roots in Aristotelian philosophy. This approach needs to be filled out with a description and evaluation of the form of reasoning followed in the biomedical decision-making process, which is provided in Chap. 7 by José-Antonio Seoane, who analyses the structure and principal characteristics of reasoning and the deliberative method in the field of bioethics, suggesting how it can be improved by incorporating elements from the legal theory and the theories of legal argumentation.

The final two chapters of the book are much more legal in nature. In Chap. 8, Juan Cianciardo studies the link between the legal principle of proportionality or reasonableness and the moral principle of double effect, classical in origin but of enormous relevance in the current ethical debate, in order to explore their validity and methodological suitability in the context of biomedical decision-making, where to all appearances contradictory human rights come into play. Finally, Vicente Bellver devotes Chap. 9 to a more institutional approach, analysing the conditions needed to ensure good bioethical deliberation through a comparison of the most significant bioethical committees on the international stage, pointing out their respective strengths and weaknesses.

Logroño, Spain
May 2016

Pedro Serna

Contents

1. **Principlism in Bioethics** 1
   Tom L. Beauchamp

2. **Values and Bioethics** 17
   Diego Gracia

3. **A Human Rights Approach to Bioethics** 31
   Roberto Andorno

4. **Philosophical Imperialism? A Critical View of North American Principlist Bioethics** 43
   Carolina Pereira-Sáez

5. **Principlism and Normative Systems** 57
   Óscar Vergara

6. **Types of Action and Criteria for Individualizing Them: The Case of Omission of Life-Saving Care** 73
   Pilar Zambrano

7. **Bioethics, Deliberation and Argumentation** 89
   José-Antonio Seoane

8. **The Principle of Proportionality, Rights Theory and the Double Effect Doctrine** 107
   Juan Cianciardo

9. **International Bioethics Committees: Conditions for a Good Deliberation** 127
   Vicente Bellver

Chapter 1
Principlism in Bioethics

Tom L. Beauchamp

The theoretical and practical roles of moral principles in principlist theory is the subject of this chapter. I start by discussing the historical background of the emergence of basic universal principles in bioethics. I then analyze the nature of the moral commitments in the universal framework of principles that I have developed over the last 40 years with James Childress. I also discuss how universal principles are fashioned into particular moralities such as those found in professional medical ethics, and the circumstances under which particular moralities are consistent with universal morality. Finally, I show the relevance of principles for global research ethics and for an understanding of the rights of minorities in multicultural societies.

1.1 The History and Origins of Principles in Bioethics

Despite the centuries-old history of medical ethics, the idea of a framework of basic moral principles at the core of medical ethics and research ethics has a short history that begins only in the later decades of the twentieth century.

T.L. Beauchamp (✉)
Department of Philosophy, Georgetown University, Washington, DC, USA
e-mail: beauchat@georgetown.edu

© Springer International Publishing Switzerland 2016
P. Serna, J.-A. Seoane (eds.), *Bioethical Decision Making and Argumentation*,
International Library of Ethics, Law, and the New Medicine 70,
DOI 10.1007/978-3-319-43419-3_1

1.1.1 *The Hippocratic Tradition's Lack of Principles*

Basic principles never played a significant role in the history of physician medical ethics, despite the near reverence that has attended the famous Hippocratic maxim, 'Above all [or first] do no harm'. The Hippocratic Oath itself was more or less a series of religious vows from ancient Greece, and these vows did not take the form of principles. The Hippocratic tradition neglected almost all of the problems of truthfulness, privacy, justice, communal responsibility, the vulnerability of research subjects, and the like that bother us today.

1.1.2 *The Virtues in Percival's* **Medical Ethics**

Historically, the health care ethics outlook in Europe and America has been one largely of maximizing medical benefits and minimizing risks of harm and disease. The finest work in professional medical ethics prior to the twentieth century was British physician Thomas Percival's 1803 book *Medical Ethics*. This book was influential in Great Britain and the United States, and set the terms of medical ethics for the next 150 years. However, Percival's book was not about basic principles. He was interested in the virtues of the physician, which constituted his moral framework. Basic principles were nowhere to be found.

1.1.3 *The Principles in the Nuremberg Code (1947)*

In 1946–1947, the American Medical Association, through its representative Andrew Ivy, testified at the Nuremberg Medical Trials that the Association had a well-established research ethics of principles. However, this claim was untrue. The principles in the Nuremberg Code had been drawn up by Ivy specifically for the trials at Nuremberg. The principles Introduced by the AMA and by the Nuremberg Code had no prior history. These principles also had no significant historical influence after Nuremberg. The principles were never put into effect in any country or by any institution. East and West Germany rejected them, and no one in the United States took them seriously as a code for U.S. practice.

1.1.4 *The Principles in the Declaration of Helsinki (1964)*

By contrast, the 1964 Declaration of Helsinki of the World Medical Association and its later revisions have been quite influential. While the Declaration is specifically set out in terms of what it calls 'principles', its norms were originally, and correctly, described by the WMA itself as 'recommendations as a guide to every doctor in biomedical research involving human subjects' (today the wording is simply 'the Declaration is addressed primarily to physicians'[1]). This description is accurate. Helsinki has never truly been a document that presents basic ethical principles of obligation; it is a body of descriptions and norms grounded in experience and consensus. It presents practical ethical guidelines for research investigators and their institutions. No basic moral principle is presented.

1.1.5 Frameworks of Basic Principles in the 1970s

Basic principles that can be understood with relative ease by the members of various disciplines first came to prominence in some 1970s developments in what was just beginning to be called 'bioethics'. As some now notorious scandals in biomedical research were being assessed in the 1970s, frameworks of basic, general principles of ethics were developed in a manner that allowed them to be readily understood by people with diverse forms of professional training and from many traditions of moral practice.

Two primary sources developed a framework of basic principles. One was the *Belmont Report* (and related documents) of the U.S. National Commission for the Protection of Human Subjects of Biomedical and Behavioral Research (National Commission 1978). The other source was *Principles of Biomedical Ethics*, which I co-authored with James F. Childress (Beauchamp and Childress 1979; 7th edition 2013). The goal of the former was a general statement of principles of research ethics, whereas the goal of latter was to develop a set of general principles suitable for biomedical ethics more generally, including ethical problems in medicine, research, nursing, and public health.

1. **The *Belmont Report* (1979)**. The *Belmont Report* was published in the *Federal Register* in 1979 as a set of basic moral principles that would justify corresponding practical guidelines in U.S. federal regulations. The purpose of the *Belmont Report* was to ensure that basic principles would become embedded in the United States research oversight system so that meaningful protection was afforded to all research participants. These principles did ultimately become entrenched in government guidelines in the U.S., where they have formed an influential and enduring basic framework for analyzing ethical issues that arise in clinical research.
2. ***Principles of Biomedical Ethics* (1979)**. The book *Principles of Biomedical Ethics* was being written at the same time the *Belmont Report* was underway. One of the proposals in this book was that medicine's traditional preoccupation

---

[1] World Medical Association (2008, A.2). Today the Declaration is best interpreted in terms of what is set out below as a set of specification of universal moral principles, though no such principles are invoked or even mentioned in the Declaration.

with a beneficence-and-care-based model of physician ethics be augmented by a principle of respect for the autonomy of patients and by wider concerns of social justice. This framework has been under discussion in many countries since its initial publication.

These two works are historically the first—and remain today the only influential—publications of a framework of basic principles of bioethics. Only the second has been regarded as a form of principlism, and only the second will be discussed below.

1.2 The Four-Principles Framework

The concept of a basic principle is analyzed in principlist theory as an abstract moral statement of obligation that is one part of a framework of essential starting points in the landscape of the moral life. If a basic principle were dropped from this framework, the moral life would change dramatically. Ethics would no longer be what we know it to be if it lacked even a single basic principle, just as a landscape would not be the same landscape if its most prominent rocks, trees, or plants were removed from it. Specific rules for health care ethics can be formulated by reference to these general principles, but neither specific rules nor practical judgments can be straightforwardly deduced from the principles.

1.2.1 Principles as Nonabsolute

All basic principles can in some contexts be justifiably overridden by other moral norms with which they come into contingent conflict. For example, we might justifiably disclose confidential information about a person to protect the rights of another person. Principles, duties, and rights (even human rights) are not absolute. That is, they are imperatives, but not categorical imperatives. In ethics, as in law, there is no escape from exercises of judgment in using principles whenever there are conflicts between the basic principles themselves.

A hierarchy of general principles cannot be constructed so that one can be confident that one principle will always take precedence over all principles. No principle always has primacy. However, in some cases, including codes of professional ethics, we may need a structured moral system of guidelines in which a certain class of rights or principles does in that context have a fixed priority over others—e.g. we might attach priorities to certain rights in the constitutions of a political state. However, no moral theory or professional code of ethics has ever successfully presented an ordered priority of moral principles so that the principles are entirely free of conflicts, exceptions, and the need to exercise judgment in balancing more than one moral demand.

1.2.2 *The Basic Principles in the Principlist Framework*

The principles in the principlist framework are grouped under four general categories: (1) respect for autonomy, (2) nonmaleficence, (3) beneficence, and (4) justice. The choice of these four principles of obligation as the framework for moral decision-making in bioethics derives, in part, from professional roles and centuries of tradition in medical ethics. Nonmaleficence and beneficence have played important roles in the history of medical ethics, whereas respect for autonomy and justice were neglected in this history and have risen to prominence only recently. All four types of principle are needed to provide a comprehensive framework for biomedical ethics, but this framework is abstract and thin in content until it has been further specified—that is, interpreted and adapted for particular circumstances—a subject to which I return later.

I concentrate in this section on the meaning and moral commitments of each of these basic principles and on the nature of the framework as a unit.

1.2.2.1 Respect for Autonomy

The principle of respect for autonomy is rooted in the idea that personal autonomy is self-rule free of *controlling* interferences by others. The two indispensable conditions of autonomy are *liberty* (the absence of controlling influences) and *agency* (self-initiated intentional action). Each of these two conditions is indeterminate until developed in a conceptual analysis of, or theory of, the notions of 'respect' and 'autonomy'.

The principle of respect for autonomy is analyzed in principlism as stating both a negative obligation and a positive obligation. As a negative obligation, autonomous actions should not be subjected to controlling constraints by others. As a positive obligation, this principle requires respectful and appropriate informational exchanges and actions that encourage and foster autonomous choice. In health care and research involving human subjects respect for autonomy obligates professionals to disclose information, to receive an informed consent to significant interventions, to probe for and ensure understanding and voluntariness, and to avoid the manipulation of patients and subjects. True respect therefore requires more than mere noninterference with decision-making and action.

This basic principle has often been misrepresented in publications about principlism as a principle of individualism, but this characterization is incorrect. Individualism is the unacceptable theory that the interests of individuals are ethically supreme over those of communities. Nothing is more antithetical to morality than individualism, and the four-principles approach wholly rejects it. A related misunderstanding holds that principlism prioritizes the principle of respect for autonomy over other principles and moral rights. The principlist theory rejects this conception as well. The principle of respect for autonomy has no form of priority in biomedical ethics other than being one *basic* principle among the others in the framework.

1.2.2.2 Nonmaleficence

The principle of nonmaleficence requires abstention from causing harm to others. This principle has commonly been presented in Medicine as the injunction: 'Above all do no harm'. Some commentators say that the principle of beneficence implies that we act beneficently by preventing harm and that therefore beneficence is the same principle as nonmaleficence. But in principlism nonmaleficence requires *not acting*—that is abstaining from acting so as not the cause harm—whereas beneficence requires *acting*—in particular acting to benefit others, for example by acting to remove harm-causing conditions. Thus, beneficence cannot involve or entail nonmaleficence because they are wholly different principles—the one of avoiding taking actions and the other of taking actions.

The term 'harm' in discussions of the principle of nonmaleficence does not entail wrongful injuring or maleficence. 'Harm' refers to a thwarting, defeating, or setting back of the interests of an individual, whether caused intentionally or unintentionally. The term 'interest' does not here refer to what a particular being is interested in, seeks, or desires. It refers to that which is in an individual's interest—a welfare condition or welfare advantage. An action by one party that harmfully affects another's interests may be fully justified and in no way wrong, maleficent, or unjustified. For example, there are justified amputations of a patient's leg, justified punishments of physicians for incompetence or negligence, etc. Causing harm may be right or wrong, but rightness or wrongness cannot be determined merely by the fact that harming occurred. Individuals can be harmed without any action by another. For example, individuals can be harmed through disease, natural disasters, and bad luck.

Numerous problems of nonmaleficence have been found in the histories of health care and research, some involving blatant abuses of persons and others involving subtle or unresolved questions. Blatant examples of failures to act nonmaleficently are found in physicians' classifications of political dissidents as mentally ill, thereafter treating them with harmful drugs and incarcerating them with insane and violent persons (Bloch and Reddaway 1984, ch. 1). More subtle examples are found in the indiscriminate use of medications for the treatment of aggressive and destructive patients. These common treatment modalities are helpful to many patients, but unjustifiably harmful to others.

1.2.2.3 Beneficence

A principle or rule of beneficence is a statement of a moral obligation to act for another's benefit, helping the other to further his or her important and legitimate interests. No moral demand placed on physicians is traditionally more important than beneficence in the care of patients. Beneficence is a foundational value in all health care ethics. Many specific duties in medicine, nursing, public health, and research are expressed in terms of a positive obligation of beneficence to come to the assistance of those in need of treatment or in danger of injury.

Principles of beneficence in medical ethics require, among other things, that we prevent harms and heal when harm has occurred. These harms are the pain, suffering, or disability caused by injury and disease. The physician who professes to 'do no harm' is not usually pledging merely to not cause harm, but to strive to create a positive balance of goods over risks of harms during the course of interventions. Persons engaged in medical practice, research, and public health know that risks of harm presented by interventions often must be weighed against possible benefits for patients, subjects, and populations.

Many acts of beneficence are obligatory. However, some beneficent acts are performed from nonobligatory, optional moral ideals, which are standards that belong to a morality of meritorious aspiration in which individuals or institutions adopt goals and practices that are not obligatory for everyone. It has been difficult to show where the line is to be drawn between obligatory and nonobligatory beneficence, and disagreements persist in moral theory regarding how much is demanded by obligations of beneficence. However, role responsibilities in medical practice contexts are widely, and correctly, regarded as major forms of obligatory beneficence.

A number of controversial issues in biomedical ethics concern how public policy could, and should, change if obligations of social beneficence were given more strength in policy formulation than they have traditionally been afforded. The foundations of public policy regarding organ procurement provide an instructive example. Established legal and policy precedents in many countries require express consent by a decedent before death or by the family after death. A near absolute right of autonomy to decide about the disposition of organs and tissues has been the prevailing norm. However, this approach impairs the efficient collection of needed tissues and organs, and many people die as a result of the shortage of organs. The scarcity of organs and tissues and the inefficiency of the system have prompted proposals for reform of the current system of procurement, with the goal of creating more space for social beneficence—a goal principlists have generally supported, while exercising caution in determining how much autonomy to protect and how much to give up in deference to beneficence.

1.2.2.4 Justice

Finally, we come to principles of justice. Every civilized society is a cooperative venture structured by moral rules of justice that define the terms of cooperation. A person has been treated justly if treated according to what is fair, due, or owed. A prime example of the need for general principles of justice is the need to distribute health care and its costs fairly. Some governments, especially the United States, tend to pay for many useless procedures—a waste of resources that may deprive others in society of adequate health care. A basic ethical problem in every society is how to structure a principled system such that burdens and benefits are fairly and efficiently distributed and a threshold condition of equitable levels of health and access to health care is in place.

Numerous issues in health care ethics in recent decades center on special levels of protection and aid for vulnerable and disadvantaged parties in health care systems. These issues range across clinical ethics, public health ethics, and research ethics. The four-principles approach deals with several of these issues without proposing a single unified theory of justice. For example, principlists address issues in research ethics about whether research is permissible with groups who have been repeatedly used as research subjects, though the advantages of research are calculated to benefit all in society. They argue that since medical research is a social enterprise for the public good, it must be accomplished in a broadly inclusive and participatory way.

Some of the most important issues in the ethics of health policy today are issues of social justice. The main moral problem in global ethics is how to structure both the global order and national systems that affect health so that undue burdens are avoided, appropriate benefits are secured, and services are fairly distributed using a threshold condition of equitable levels of health and access to health care. Globalization has brought a realization that problems of protecting health and providing beneficial services are international in nature and that their alleviation will require a restructuring of the global system. As principlists recognize, principles of beneficence and justice are appropriately joined together in many of these discussions.

The basic commitments of justice found in *Principles of Biomedical Ethics* are modeled on egalitarian social justice. It demands creation of a just system of healthcare for both national and international systems of distribution of public health services and healthcare goods. Limits on egalitarian distributions are of essential, but the subject of how to set limits cannot be considered here.

1.3 The Centrality of Common Morality in Principlist Theory

Common morality theory is a vital part of the principlist account of biomedical ethics. Throughout human history we have learned that the human condition tends to deteriorate into misery, confusion, violence, and distrust unless certain moral principles are enforced through a public system of norms. The maintenance and enforcement of basic moral standards such as not lying, not stealing others' property, keeping promises, respecting the rights of others, and not killing or causing harm to others are necessary conditions of a decent life and the reason why a common morality exists.

Why have some principles become central, basic parts of a common morality, whereas other norms have not? The answer is that these principles are requisite for the promotion of human flourishing. Maintenance of these principles prevents or limits problems of indifference, conflict, suffering, hostility, scarce resources, limited information, and the like. These norms may not be necessary for the survival of a society (cf. Bok 1995, 13–23, 50–59), but they are necessary to ameliorate or counteract the tendency for the quality of people's lives to worsen or for social relationships to disintegrate (Warnock 1971, 15–26).

The universal character of the human experience and of social responses to threatening conditions helps *explain* why there is a common morality, but it does not *justify* the principles. What justifies them is that they are the norms best suited to achieve the objective of morality, which is the promotion of human flourishing by counteracting human circumstances in interactions with others that cause the quality of people's lives to worsen. Once the objectives of morality have been identified, a set of standards is justified if and only if it is the best means to the end identified when all factors—including human limitations, shortcomings, and vulnerabilities—are taken into consideration. If one set of norms will better serve the objective of morality than a set currently in place, then the former should displace the latter (Beauchamp 2010).

Some commentators think that Childress and I hold that the set of four principles constitutes the full set of universal norms. However, we merely select four principles from the larger set of principles in the common morality for the purpose of constructing a normative framework *for biomedical ethics*.

We also understand the common morality as comprised of *virtues*, *ideals*, and *rights*. Examples of universal virtues, or moral character traits, include honesty, integrity, nonmalevolence, trustworthiness, and truthfulness. These virtues are universally admired traits (Nussbaum 1988), 33–34, 46–50). A person is deficient in moral character if he or she lacks one or more of these traits. Examples of universally praised ideals include charitable goals, community service, mentoring professional colleagues, dedication to one's job that exceeds obligatory levels, and service to the poor. These aspirations are not moral obligations, and not morally *required* of persons, but they are universally *admired* and *praised* in persons who accept and act on them (Gert 2004, 20–26, 76–77). Finally, universal rights (human rights) are vital parts of the common morality. Rights are justified claims to something that individuals or groups can legitimately assert against other individuals or groups. Human rights, in particular, are rights that all humans possess (Feinberg 1980, 139–60). Human rights cross national and cultural boundaries with standards that transcend unacceptable norms and practices.

Principlist theory defends a strict version of the thesis that rights and obligations are correlative. Basic principles and human rights are indissolubly bound together: Universal obligations uniformly entail corresponding rights. These obligations must be bona fide universal moral obligations, not merely self-assumed obligations or personal moral ideals, such as 'obligations' of charitable giving. For example, the right to have information remain confidential after a confidential disclosure is correlative to obligations to protect confidential information. Every true obligation is correlative to someone's right—though it would be overreaching to say that all of these rights are *human* rights, a concept in need of more careful treatment than it has yet received in literature on the subject.

Summing up, the common morality is comprised of a universal set of moral norms. The term 'common morality' references the entire moral system of universal standards: principles of obligation, virtues, rights, and moral ideals. 'Principles of biomedical ethics' refers only to a subset of common morality.

1.4 Specifying Principles to Render Them Practical

General principles of ethics are not sufficient by themselves to determine the content of specific moral judgments. Too little content is found in abstract principles to determine the many moral rules and judgments needed in the moral life. All abstract norms therefore must be given additional content in order to achieve practical guidance about how much information must be disclosed, how to maintain confidentiality, when and how to obtain an informed consent, and the like.

1.4.1 The Method of Specification

Giving specific content to general principles and rules is accomplished by narrowing their scope (whether the norms were previously unspecified or were norms that had previously been specified). Specification occurs by 'spelling out where, when, why, how, by what means, to whom, or by whom an action is to be done or avoided' (Richardson 1990, 289). For example, one possible specification of 'respect for autonomy' is 'respect the autonomy of competent patients when they become incompetent by following their advance directives'. This specification will work well in some medical contexts, but will not be adequate in others where it will need additional specification. Progressive specification can continue indefinitely, gradually reducing conflicts and uncertainties that abstract principles themselves cannot resolve. This is how professional codes of ethics, hospital ethics committees, and public policies can be progressively made practical and clear in institutional contexts.

More than one line of specification of principles is commonly available when confronting practical problems and moral disagreements. Different persons or groups may for good reason offer conflicting specifications, thus potentially creating multiple particular moralities. For deeply problematic issues such as abortion, animal research, aid for disaster relief, health inequities, and euthanasia, competing specifications will be offered.

1.4.2 Justifying Specifications Using a Method of Coherence

Since numerous specifications may be available, questions arise about how to choose between alternative specifications and how to justify one's choice. A specification is justified, in the principlist account, if and only if it is consistent with the norms of common morality and maximizes the coherence of the overall set of relevant, justified beliefs of the party doing the specification. These beliefs could include empirically justified beliefs, justified basic moral beliefs, and previously justified specifications. This position is a version of the philosophical account of justification and theory-construction in ethics known as wide reflective equilibrium, a view based on the theory of reflective equilibrium in the philosophy of John Rawls (Daniels 1996, 96–114). This theory holds that justification in ethics occurs through a reflective testing of moral beliefs, moral principles, and theoretical postulates with the goal of making them as coherent as possible. The goal of any given specification is to achieve an equilibrium while also resolving a contingent conflict of principles.

The goal of a coherence of norms assumes that some starting norms are available to be shown coherent. One must start in moral reflection with a particular body of beliefs that are acceptable initially without argumentative support. According to Rawls' account, which principlists adopt and adapt for use with general principles, the method of reflective equilibrium begins with what Rawls calls considered judgments. These moral and political convictions are those in which we have the highest confidence and believe to contain the lowest level of bias or distortion of perspective. Considered judgments deserve this status because they are so deeply entrenched in moral thinking that any morally decent person would accept and act on them. Examples of the sort cited by Rawls are norms about the wrongness of racial and sexual discrimination, religious intolerance, and political repression. But considered judgments occur at all levels in moral thinking from those made in particular situations (e.g. a compassionate and caring response when a person is ill) to basic moral principles and rights (e.g., human rights). In principlism, the basic principles constituting the moral framework are themselves considered judgments with which one starts in moral reflection and are acceptable initially without argumentative support.

The process of achieving a state of equilibrium through moral deliberation is a process of reflective testing in which one prunes, adjusts, and specifies one's considered judgments and beliefs to render the whole system of beliefs coherent. The resultant moral and political norms can then be tested in a wide variety of previously unexamined circumstances (e.g., in circumstances of apparent conflicts of interest never previously examined) to see if incoherent results emerge. If incoherence arises, conflicting norms must be adjusted further to the point of coherence.

Achieving a state of reflective equilibrium in which all beliefs fit together coherently, with no residual conflicts or incoherence is a vital goal, but it will never be comprehensively realized because it is and will always remain an ongoing project. There is no reason to expect that the process of rendering moral norms coherent by specification will come to an end or be perfected. Nonetheless, this ideal is not merely a utopian goal toward which no progress can be made. Particular moralities of individuals and groups are a continuous process of improvement rather than a finished product.

To take an example from the ethics of the distribution of organs for transplantation, imagine that an institution has used and continues to be attracted to two policies, each of which rests on a basic rule: (1) distribute organs by expected number of years of survival (to maximize the beneficial outcome of the procedure), and (2) distribute organs by using a waiting list (to give every candidate an equal opportunity). These two distributive principles are inconsistent and must be brought into equilibrium in the institution's policies if both are to be retained. Both can be retained in a system of fair distribution if coherent limits are placed on the norms. For example, organs could be distributed by expected years of survival to persons 65 years of age and older, and organs could be distributed by a waiting list for 64 years of age and younger. Proponents of this policy would need to justify and render as specific as possible their reasons for these two different, but incoherent, commitments. Such proposals would have to be made internally coherent in the system of distribution and also would need to be made coherent with all other principles and rules pertaining to distribution, such as norms regarding discrimination against the elderly and fair payment schemes for expensive medical procedures.

1.5 Particular Moralities and Specification

Particular moralities are created by processes of specification. These moralities include the many responsibilities, ideals, attitudes, and sensitivities found in, for example, cultural traditions, religious ethics, and professional guidelines. Particular moralities can differ extensively, and justifiably, in the content of their beliefs, their emphases, and their practice standards. One society might emphasize the liberty of individuals, while another emphasizes justice in the distribution of social goods over liberty rights. What is unacceptable in one society might be justifiably condoned in another. For example, certain forms of clothing and/or nudity might be acceptable in one society but unacceptable in another.

The reason why norms in particular moralities often differ is that the universal starting points in the common morality—its basic principles—can be legitimately specified in different ways to create different guidelines and procedures. These differences are acceptable as long as basic principles are not violated and there is a serious attempt at justification of the specifications, as discussed above.

Good examples of particular moralities that contain at least some specifications not found in other moralities are *professional moralities* such as those in biomedical research, medical practice, nursing practice, veterinary practice, and the like. Their moral codes, declarations, and standards of practice often legitimately vary from codes in other countries or jurisdictions in the ways they handle justice in access to health care, justified waivers of informed consent, permissible relations to government officials, conflict of interest, privacy provisions, confidentiality protections, and the like.

It has sometimes been proposed that a framework of moral rules for specific fields such as psychiatric ethics needs rules specifically designed for the problems in that field—e.g., principles specifically designed for problems of psychiatry pertaining to competence, informed consent, confidentiality, self-harm, vulnerability, criminal responsibility, involuntary civil commitment and compelled treatment after a patient's refusal of treatment, and use of drugs to render a patient competent to stand trial. The claim is that the distinctive character of psychiatry requires a unique ethics (Radden 2002, 52).

This assessment is ill-advised if it means an abandonment of or irrelevance of universal principles. It is certainly correct that each professional field needs *specifications* of universal principles that are different than specifications needed in other fields; but the general principles in the principlist framework are relevant for, and required in, all biomedical fields. To make principles and rules entirely relative to disciplines or fields invites confusion and a loss of universal validity.

Other examples of particular moralities that contain differing specifications are found in *religious moralities*. Religious traditions may have multiple moralities within the spread of a single religion. Protestant Christianity is an example. Each Protestant sect may have its own body of rules of ethics—e.g., rules about abortion, blood transfusion, cosmetic surgery, cessation of life-sustaining treatment, and organ transplantation. These different Christian groups *share* universal moral principles, but they may not share certain specific moral rules that contribute to making each one distinctively the religious group it is in its moral demands. As with any particular morality distinctive to a tradition, a religious group will state what is permissible and impermissible, what is obligatory and nonobligatory, what has moral status and what does not, etc.

Many significant advantages are resident in well-specified particular moralities. Universal morality—being general and only general—does not have and cannot have the same richness and specificity that a particular morality does. Moreover, a particular morality does not forego universal morality; it retains and is governed by universal morality. These are great advantages for particular moralities.

Consider as an example of specifications in particular moralities rules pertaining to permission for a necessary medical intervention when a patient does not have capacity to consent. A common provision is that a legal guardian must give consent. But who counts as a legal guardian and is there an order of guardianship if a first guardian refuses to serve in the role? If a guardian refuses a medically necessary intervention, is this refusal decisive? Such questions about required permission, guardian selection and authority, and valid surrogate informed consent must be determined suitable to specific traditions with rules that specify universal principles.

1.6 The Idea of Eastern and Western Moralities

To extend the discussion of particular moralities, consider the idea that there are *basic* moral differences between Eastern and Western cultures. Confusion continues to be plentiful about the role that universal principles play in making judgments about moral claims made in different moral traditions. Little if any real evidence supports the commonly reported thesis that the East—that is, Asia—has fundamentally different moral traditions of rights, liberty, respect for autonomy, and respect for families from those in the West—that is, Europe and the Americas.

I agree with Amartya Sen's views in 'Human Rights and Asian Values', where he argues that 'no quintessential [moral] values . . . *differentiate* Asians as a group from people in the rest of the world'. He finds that the major components of universally valid ideas of liberty and basic rights of liberty—for example, liberties in a political state—are found in both Eastern and Western traditions. The claim that these ideas are friendly to Western traditions and alien to Eastern traditions he finds 'hard to make any sense of' (Sen 1997, 10, 13, 17, 27, 30).

Principlist theory fully aligns with this perspective but does not hold that a principle such as respect for individual autonomy is given precisely the same status and prized to the same extent in all Eastern traditions as it might be in various Western cultures. Many populations in the East may prioritize community and relationships of specifiable kinds over individual autonomy and cultural independence to *a higher degree* than do certain populations in the West, but this thesis does not entail that Eastern populations deprecate or reject individual autonomy and political liberties. Nor does it indicate that Western populations deprecate community and relationships.

Many critics of principlism have said that the principlist framework derives from 'western ethics' and rests on a set of 'western moral principles'. But there is no large or unbridgeable gap between so-called eastern and western ethics—nor is there a significant gap between American and European ethics. No distinctly eastern ethics and no distinctly western ethics are to be found.

One can be both a particularist and a universalist about moral norms, a position that follows from the above discussions of common morality, particular moralities, and specification. All justified particular moralities share the norms of universal morality with all other justified particular moralities. That is, all justified particular moralities share universal morality, while acknowledging legitimate tools to refine its unclarities and to allow for additional specification of principles, rules, and rights. If local standards do not embrace universal principles, they are not justified to the extent they violate or ignore these principles.

Some writers in bioethics say that we live in a multicultural world in which diverse particular moral cultures can live together peacefully, without need for universal principles (Engelhardt 1996). However, multiculturalism, properly understood, is not a pluralism or a relativism of basic principles. It is a theory of universal principles to the effect that particular moralities are owed respect because morality itself demands that everyone's views are owed respect. The objective of multiculturalism is to protect vulnerable groups when they are threatened with marginalization and oppression caused by one or more dominant cultures. The central thesis of multiculturalism is that respect is owed to people of dissimilar but peaceful cultural traditions because it is unjust and disrespectful to marginalize, oppress, or dominate persons merely because they are of an unlike culture or subculture. The basic moral premises in multiculturalism, then, are universal-principle driven theses about rights, justice, respect, and nonoppression. Rules of respect for different cultures are justified in terms of universally valid principles of ethics.

1.7 The Global Acceptance of Universal Principles in Research Ethics

The reach of universal principles of bioethics and related rules are now evident in globally accepted rules in research ethics (in biomedical fields). Forty years ago there were no universally recognized rules of research ethics, but today we can see a vast similarity, in countries on every continent, in professional codes, laws, and regulations governing research with human subjects. Any differences in the rules of countries are insignificant in comparison to the similarities in the shared moral principles and regulatory norms governing how biomedical research can and cannot be conducted. Some of these rules were three decades ago met with skepticism in many professional societies, but today they are universally accepted in research ethics.

Here are some examples of globally accepted rules of research ethics:

Disclose all material information to subjects of research.
Obtain a voluntary, informed consent to biomedical interventions.
Receive a valid surrogate consent from a legally authorized representative for incompetent subjects.
Subjects entered in a study have a right to be informed about study results at the conclusion of the study.
Maintain secure safeguards for keeping personal information about subjects private and confidential.
The design and conduct of each human-subjects study must be set out in a research protocol.
Ethics review committees must scrutinize and approve research protocols before a study can begin.
Research cannot be conducted unless its risks and intended benefits are reasonably balanced; and risks must be reduced to avoid excessive risk.
Special justification is required if proposed research subjects are vulnerable persons.

These rules are examples of what was earlier described as specifications of basic principles. Principlists hold that each rule is grounded in and justified by one or more basic moral principles, though progressive specification is involved in each rule listed above—as can be historically traced in the expansion of research ethics in the last 40 years.

1.8 Conclusion

The several sections in this chapter move to the conclusion that a universal set of moral principles constitutes the basic core of medical morality. Although the specific normative content of each principle is thin, they stand as a wall of basic moral standards that cannot be justifiably breached in any culture or by any group or individual, even though they can sometimes be justifiably overridden when moral conflicts occur among the principles themselves. These principles give us a moral compass and a bulwark against descent into moral chaos.

References

Beauchamp, Tom L. 2010. *Standing on principles: Collected essays*. New York: Oxford University Press.
Beauchamp, Tom L., and James F. Childress. 1979 and 2013. *Principles of biomedical ethics*. 1st & 7th edn. New York: Oxford University Press.
Bloch, Sidney, and Peter Reddaway. 1984. *Soviet psychiatric abuse: The shadow over world psychiatry*. Boulder: Westview Press.
Bok, Sissela. 1995. *Common values*. Columbia: University of Missouri Press.
Daniels, Norman. 1996. Wide reflective equilibrium in practice. In *Philosophical perspectives on bioethics*, ed. L.W. Sumner and J. Boyle, 96–114. Toronto: University of Toronto Press.
Engelhardt Jr., H. Tristram. 1996. *The foundations of bioethics*, 2nd ed. New York: Oxford University Press.
Feinberg, Joel. 1980. *Rights, justice, and the bounds of liberty*. Princeton: Princeton University Press.
Gert, Bernard. 2004. *Common morality: Deciding what to do*. New York: Oxford University Press.
National Commission for the Protection of Human Subjects of Biomedical and Behavioral Research. 1978. *The Belmont report: Ethical principles and guidelines for the protection of human subjects of research*. Washington, DC: DHEW Publication OS 78-0012.
Nussbaum, Martha. 1988. Non-relative virtues: An Aristotelian approach. In *Ethical theory, character, and virtue*, ed. Peter French et al. Notre Dame: University of Notre Dame Press.
Radden, Jennifer. 2002. Notes towards a professional ethics for psychiatry. *Australian & New Zealand Journal of Psychiatry* 36(2002): 52–59.
Richardson, Henry S. 1990. Specifying norms as a way to resolve concrete ethical problems. *Philosophy and Public Affairs* 19(1990): 279–310.
Sen, Amartya. 1997. *Human rights and Asian values*. New York: Carnegie Council.
Warnock, G.J. 1971. *The object of morality*. London: Methuen & Co.
World Medical Association. 2014. Ethical principles for medical research involving human subjects. 59th WMA General Assembly, Seoul, October 2008. Available at http://www.wma.net/en/30publications/10policies/b3/17c.pdf. 12 June 2014.

Chapter 2
Values and Bioethics

Diego Gracia

2.1 A Biological Need

Since ancient times human beings seem to have had a clear idea of the biological deficit of the human species. The extreme immaturity of humans at birth, almost unparalleled in the history of mammalian species, is quite astonishing. And even when that initial deficit is overcome, their biological traits remain comparatively deficient: they are not strong as a lion, nor do they have the eyesight of a lynx, nor the speed of a greyhound, etc. From the point of view of Darwin's theory of biological evolution, this means that the human species is not "fitted to the environment" to the point that if it were not for a distinctive phenotypic trait, which we call the human mind or human intellect, human beings would long ago have disappeared from the face of the earth.

From the biological point of view, the human mind is a phenotypic trait like many others, such as hair color or muscle strength, with a primary biological function, namely to fit human beings to the environment. But it so happens that this way of fitting an organism to the environment is new, differing from all previous evolutionary pathways. In the animal world, adaptation to the environment means that the environment selects those that are the fittest to survive therein, all others being penalized as unfit. This is the process Darwin called "natural selection", because it is nature, the environment, and not animals or genes, which makes the selection. Human beings, on the other hand, possess a certain ability to take the initiative, selecting, modifying and fitting the environment to them, in other words adapting the environment to their needs. Hence, due to the specific nature of human intelligence, the "adaptation *to* the environment" characteristic of previous biological evolution is transformed into a new mechanism that we can call "adaptation *of* the

---

D. Gracia (✉)
Fundación de Ciencias de la Salud, Madrid, Spain
e-mail: dgracia@fcs.es

environment". Since it is not fitted to the environment, the human species attempts, through the vehicle of its intelligence, to adapt the environment to the human needs. The result of this transformation or humanization of nature is what we call "culture." Human beings cannot live in "nature", as animals do. They necessarily and permanently live in a more or less transformed or humanized nature, in other words, in "culture". The "natural environment" is transformed, as a result of intelligence, into a "cultural world".

All this is due to the peculiar characteristics of the human mind. All instruments or beings capable of processing information are to some extent intelligent. Animals are constantly processing information in order to interact with the environment, this being the function of the nervous system. For example, when they are running, horses need to foresee obstacles in order to avoid them. The nervous system of animals is a fantastic organ of "foresight", which through a complex system of afferent and efferent pathways enables animals to process information from the environment and adapt their movements to it. In the absence of this superb way of processing information and adapting their movements accordingly, animals would constantly fail to adapt to their environment, and this is why we say they are "adapted". Human beings are also animals, but with a fundamental difference. The "forecast", which in the case of a galloping horse is automatic, fitting it to the environment, is open in human beings, and can only be achieved by means of a "project". Human beings have the ability to "project" their actions, an ability they possess not only because they are "forced" to do so by purely biological needs, but also because it is only through projects that human beings can change the environment, cultivating and humanizing it.

2.2 The Project Structure

Let us look more closely at the structure of a human project. The goal of each human project is to take a decision, to do something or not. What we project is always the future: we cannot project either the past or the present. Due to the projecting nature of our mind, we as human beings live in constant mental anticipation of the future, previewing it. And because the end goal of projects is to take decisions, we, as human beings, become responsible for the decisions we take. In other words, we are responsible for our own projects, and herein lies the origin of the moral life. However, not all human acts are moral, because not all of them are projected. Many acts, in human beings as well as in animals, are automatic, reflex, unconscious, instantaneous, etc., but we are only morally responsible for our planned actions.

Project and responsibility exist in a reciprocal relationship. What we project are ends, the goals of our actions, and these ends return to us, making us responsible for them. In a sense a project can be compared to a boomerang, which we launch forwards and which then comes back to us in the form of responsibility. We take responsibility for the project: we are its end, and therefore the end of the projected ends.

This is what Kant called being "end in itself", the definition of the moral subject, unlike the purely natural objects (Kant 1968, 428).

Projecting is a complex phenomenon, one which brings into play the full resources of our human mind. Projects must be based on certain objective data, generally called "facts", something that is self-evident and deserves no further discussion. What is surprising is that human projects never deal only with facts: in addition, they always and necessarily include "values". The reason for this is that we are unable to perceive, imagine, think, or remember something without immediately valuing it. We necessarily value it as more or less beautiful, more or less useful, etc. Facts do not exist independently from values, and all human projects include both, although we are often not aware of this. Just as the world of facts is familiar and close to us, we see the world of values as strange, dark, difficult and problematic. Facts give us security, while the presence of values makes us feel uncomfortable, filling us with uncertainty, which in turn generates anxiety, a feeling that immediately triggers what Freud called the ego defense mechanisms, first and foremost the most primitive one of all, denial. This is what makes values dangerous and uneasy, because we do not know what they are or how to handle them. Nevertheless, they are an essential part of human life and a constituent element of any project.

Facts and values are articulated in each human project in order to make a decision, constituting the third stage of the project. This process has a cognitive moment, which gives us the "facts" to be taken into account, and an emotional one, which will make the project more or less attractive and "valuable", more or less appreciated, more or less beautiful, useful, etc. But these two elements alone do not suffice to generate an action. A third operational or practical step is required, which leads us to make the decision to do or not to do something. In traditional psychology, these three stages of the project were carried out by the three faculties of the soul, intelligence, feeling and volition or will, respectively. The "facts" identified in the first will lead us to "value" them positively or negatively, and finally, at the third stage, to take the decision and perform the action or not. If the assessment is positive, the conclusion will be that we "should" perform it: this is the origin of the idea of "duty". Our first and most basic duty is to increase the value of things, promoting positive values and avoiding doing harm, or at least doing as little harm as possible, in order to perfect and humanize nature. We all seek to achieve this goal, and that is the reason why there is a tax on production and labor called value added tax. We add value through our actions. When farmers cultivate land, they are increasing its value, just as when a builder builds a house, or an artist paints a picture.

2.3 Facts and Values

The darkness of the world of values contrasts with the clarity of the world of facts. Our culture, especially from the eighteenth century onwards, revolves around so-called facts, particularly "scientific facts" or "experimental facts", which Comte classified in the mid-nineteenth century as "positive." Scientific facts are characterized by having been tried and tested through the scientific method, which makes them reliable. Founder of the positivist movement, Auguste Comte tried to organize all human life around the idea of fact. Previously, mankind had passed through two stages, the mythical and the speculative, in which they attempted to solve problems through inadequate methods, these being unbridled imagination, in the first case, and metaphysical speculation, in the second. Comte believed the time was near when the lives of all human beings would be organized around the concept of objective or scientific fact, thus solving the real problems of mankind for the first time in history. As Thomas Gradgrind said to the schoolteacher at the beginning of Charles Dickens' novel *Hard Times*: "Now, what I want is, Facts. Teach these boys and girls nothing but Facts. Facts alone are wanted in life. Plant nothing else, and root out everything else. You can only form the minds of reasoning animals upon Facts: nothing else will ever be of any service to them. This is the principle on which I bring up my own children, and this is the principle on which I bring up these children! Stick to Facts, sir." So begins the first chapter of the novel, significantly entitled: "The one thing needful." All others are superfluous. "In this life, we want nothing but Facts, sir; nothing but Facts!" (Dickens 1854, I 1)

However, the fact culture has failed to achieve its goal of solving the problems of mankind. Although it has been capable of solving or at least alleviating some of them, it has also aggravated others. For example, the First World War was seen by many European intellectuals as the failure of the type of mentality that put its trust in science and technology. Hence, from then onwards, many critical voices began to make themselves heard against this attempt to reduce human projects to just one of its constituent aspects, overlooking the second one, because the lack of attention paid to matters of value was probably the reason for that failure. These voices became stronger after the Second World War, when what we now commonly refer to as weapons of mass destruction were first developed and used, and people realized that facts without values could only lead to disaster. This realization drove Van Rensselaer Potter to coin the term "bioethics", conceived as a bridge between the new "facts" of life sciences, on one hand, and the "values" these facts put at stake, on the other (Potter 1971). Not everything that is technically possible is ethically correct, and the spectacular development of bioethics since then testifies to the increased sensitivity of human beings towards problems of this kind. The making of projects that only include facts does not mean that we can avoid values, because values are unavoidable. The consequence of doing so is that values are also included, but uncritically and unthinkingly, with the danger of including the wrong ones, and with no clear awareness of what we are doing.

2.4 The World of Values

The type of mentality described in the previous paragraph is nowadays often referred to as "instrumental rationality" and is the model or paradigm of rationality specific to the technical world. Instruments are accorded the status of "means", related to "ends."

These ends are always values, and instruments are thus at the service of values such as health, life, welfare, happiness, etc. These instruments or technical tools are also valuable, but their value is merely instrumental, because it depends on the value end to which they are related. These other values, however, cannot also be instrumental: in fact they are called "intrinsic values" or "values in themselves". So in addition to instrumental values, we have intrinsic values. In this context nothing metaphysical is meant by the term "intrinsic", which refers to those qualities or things that are valuable in themselves, unrelated to others. Pharmaceuticals have value if they cure diseases or relieve symptoms. Otherwise, we would say they are "worthless": their condition is purely instrumental, as a means to promote other values such as life, health or welfare. We can ask whether these are intrinsic values, but to answer that question, we must think whether we consider them to be valuable in themselves, independently of all other things. For example, if everything stayed the same in the world as now, but without beauty, and we consider that something important has been lost, we can conclude that beauty is valuable in itself, and is thus an intrinsic value. The same is true of many other values such as life, health, pleasure, wellbeing, peace, justice, solidarity, etc., all of which are, therefore, intrinsic values or values in themselves. At the very beginning of the twentieth century, G.E. Moore asked about "the method which must be employed in order to decide the question 'What things have intrinsic value, and in what degree?'" and he answered: "In order to arrive at a correct decision on the first part of this question, it is necessary to consider what things are such that, if they existed *by themselves*, in absolute isolation, we should yet judge their existence to be good; and, in order to decide upon the relative *degrees* of value of different things, we must similarly consider what comparative value seems to attach to the isolated existence of each." (Moore 1994, 236)

We can now return to the concept of instrumental rationality, in which all values are taken and managed as merely instrumental. The existence of intrinsic values is not accepted, something which has very important practical implications, since instrumental values have several characteristics that are not only different but opposed to those of the intrinsic ones, of which two in particular stand out over and above the rest. The first is that instrumental values are always measured in monetary units. This is because money is the purest instrumental value, the instrument of all other instruments, because it is good for nothing except to measure the value of all other instruments. So in an age like our own, in which all values are taken as instrumental, the rise of economic value to first place in the consideration of human beings seems entirely logical, finding its expression in popular or conventional wisdom in the form of sentences such as "You are what you own" or " Things are only worth what someone else is willing to pay", etc.

The second feature of instrumental values is that they are interchangeable. If I take a drug and find another with the same pharmacological effect or with fewer side effects, at a lower price, I can substitute the latter for the former without any problem. This is, again, because these items have no value in themselves. What makes them valuable in this case is health.

Intrinsic values, on the other hand, possess exactly the two opposite characteristics: they are not interchangeable with each other, and they cannot be measured in monetary units. Human beings are endowed with an intrinsic value called dignity, which make them worthy in themselves, so they can neither be exchanged nor bought or sold in the marketplace. The value of human beings cannot be measured in monetary units. Strangely enough, some things cannot be measured with the typical yardsticks of economics, usually referred to by economists as "intangible things". Dignity can be one of these, but there are others that bear no direct relationship to human beings. Velazquez and Goya, for example, were two artistic geniuses because they were capable of expressing some traits of the intrinsic value of beauty through their brushstrokes. The pictures of both artists are beautiful, but at the same time different, so they cannot be exchanged one for another. If the beauty Velázquez's paintings were to vanish from the face of the earth we would say that something valuable in itself has been lost, because that particular beauty is not replaceable by any other. With this we have introduced a new word that defines very well the difference between these two types of values: instrumental values can be "replaceable", while intrinsic ones are "irreplaceable". Each loss of intrinsic value is a small or large "tragedy."

2.5 Valuing and Values

Valuation is a mental phenomenon like many others. The human mind performs a large number of very different functions such as perceiving, imagining, remembering, thinking, feeling or wanting; another of such functions is the one that has received the names of estimating, assessing or valuing. Valuing is a subjective phenomenon, which takes place inside our minds. For example, when we see a person we immediately assess her or him as being attractive or ugly, or we assess the price of things as being expensive or cheap. This intrinsic mental activity can be called "valuation". Before he started painting The Toilet of Venus, Velazquez had to imagine and estimate its beauty. That's what led him to paint the picture. And once he did, this beauty went from being a subjective activity that took place inside the mind of Velázquez to becoming an objective reality in a picture. If the former was a phenomenon of "valuation", the latter is something entirely different and of a much more objective nature; it is a "value", the aesthetic value of that picture.

Distinguishing between subjective valuing and objective value is extremely important, because through their works, humans can do nothing more than achieve, accomplish, objectify or realize values, both positive and negative. When realized, values become independent from their author and begin their own life: although Velázquez has long since died, the picture of The Toilet of Venus and its beauty are still with us. It has become part of the set of values of our society, what we call "culture", which is nothing but the set of values that the members of a society have been realizing through their actions. Hence the tremendous importance of any human act, however small it may seem. All of them realize values or disvalues, engaging them in the common set called culture. This explains why there are cultures in which certain values are prominent over others: some cultures focus on religious values, others on patriotic values, yet others on aesthetic values, etc. And there are also cultures in which the prominent values are negative, such as corruption. There are corrupt societies, just as there are others that are painstaking, hard-working, truthful, or respectful.

The sum of values we call "culture" is transmitted to the members of each new generation at birth. Society will necessarily deliver them that legacy, built up through the acts of previous generations. In Greek this process of handing over is called *paradosis*, and in Latin *tradition*, from which we derive our word "tradition", namely the set of values that people share and transmit to the following generations. These values will constitute the background to all the projects performed by their members, and therefore of all the actions they will carry out. These actions will in turn add, modify or impoverish the previous set of values, or even transform it in a revolutionary way, but necessarily starting from the legacy received. Nobody starts afresh. Adamism is impossible.

2.6 Value Conflicts

An inherent characteristic of the world of values is its controversial nature. Each value can conflict with all others, making the determination of duties highly problematic. If there is no conflict, every human being knows the decision he or she has to take, carrying out as far as possible the value at stake. In a case of injustice, justice should be promoted; in a case of war, peace; and so on.

But values can come into conflict, something that happens when there are two or more values at stake, so that if we try to carry out one of them, we can harm the other. That is why a conflict of values comes down in the end to a conflict of duties, the problem being how to determine which duty is the correct one.

A possible solution could be to order the values at stake according to their "rank" or "hierarchy". As a general rule, intrinsic values are considered to be hierarchically superior to instrumental ones. However, apart from rank values have another characteristic, referred to by some as "strength" and by others as "urgency". According to this notion, instrumental values, in spite of being inferior in rank to the intrinsic ones, are more basic, because they support all other values. This is evident in the case of economic value. Despite being hierarchically inferior, it supports all the others. Hence the twofold moral obligation: first, to realize intrinsic values, and second, to avoid harming the instrumental ones. Neither of these criteria is the only one applicable to these conflicts, nor the most important, because our first duty is not to choose one value over others, but to promote all of them, avoiding doing harm to them and attempting to implement them to the greatest possible extent.

2.7 Values and Duties

We can now systematize the way of taking moral decisions, but the question is how to determine whether a decision is morally right or wrong. We know that our first moral duty is not to choose one of the values at stake, deleting all others, but to preserve and carry out all of them to the greatest extent possible. The question, however, is how to proceed in order to achieve this goal.

Once the conflicting values have been identified, it is necessary to analyze the possible courses of action. There are conflicts without a solution, or whose solution does not depend on us, so they cannot be a moral problem. Nor can a moral problem be said to exist if there is only one possible course of action. The question arises when there are at least two possible courses of action, a situation that is often called a "dilemma": bioethics literature is full of books and articles on moral dilemmas. My opinion is that dilemmas are very rare in human life, and that the classic dilemmas that appear in books on ethics, such as the prisoner's dilemma, are artificial, clauses having been added to them that are very difficult to find in practice. The most common situations of this kind that arise in human experience are not dilemmas but "problems", conflicts that have three or more solutions, sometimes a thousand, which we have to identify.

A particularly damaging feature of the human mind is its tendency to turn problems into dilemmas, seeing only the extreme courses of action, and ignoring all the other possible intermediate courses. The extreme courses in the case of a conflict of values always coincide with choosing one of the values at stake with absolute harm to the other, and vice versa. When there are two values, A and B, an extreme course would be to carry out value A and causing absolute harm to value B, whilst the other one would do exactly the opposite. It goes without saying that extreme courses are always harmful, since if they are carried out one of the values is completely lost, and our first duty is to realize all the positive values at stake to the maximum extent of our possibilities.

The best course usually lies somewhere in the middle, because only intermediate courses try to realize all the values at stake. Intermediate courses are not usually easy to identify, because it is much easier for the human mind to see the extremes, white and black, than the many shades of gray in between. The care taken in analyzing the intermediate courses of action determines the quality of the decision to be taken. As Aristotle said, the best course is usually an intermediate one: *in medio virtus*, according to the famous Latin saying (Aristotle 1831, 1106 b 15–1107 a 26). And ethics is not content with less than the optimal course, its purpose being not to achieve the good but the optimal. Choosing any course other than the optimal is always wrong; a bad judge is one who does not take the optimal decision, and a bad surgeon is one who operates in a suboptimal way.

2.8 Deliberation as Procedure

We said at the beginning that the human mind has a strictly projective function, allowing human beings to mentally anticipate decisions and actions. Projecting requires a method, a procedure, and this method has, since the time of Aristotle, been known by the name of "deliberation." (Aristotle 1831, 1112 a 18–1113 a 13; 1142 a 32–1142 b 33) Other uses of human reason are not deliberative, and the clearest example of this is mathematics. About the fact that two and two are four, says Aristotle, there is no possible deliberation. Nor can we deliberate on the Pythagorean theorem. Theorems can be demonstrated, which means that we know their true solution, all other possible solutions being by definition false. This is the logic in which the final valences are true and false.

Demonstrative logic, however, is very rare in human life. In fact, it cannot be applied to the specific concrete situations of our everyday life, which are resistant to being solved as mathematical problems. Just as "certainty" is inherent to demonstrative logic, in daily life we frequently need to make decisions in situations of "uncertainty." The number of inputs is so large and our capacity to deal with them so poor that we are frequently unable to achieve certainty. This is necessarily the case whenever we include the most relevant circumstances of the case and the likely consequences in the decision-making process, because it is never possible to exhaust the analysis of circumstances, and much less so to foresee all the consequences. The paradigmatic example of this is the weather forecast. Despite all the technical advances in this field, it is impossible to eliminate uncertainty; Edward Lorenz has shown that certainty not only has not yet been achieved in this domain, but that it never will be (Lorenz 1993, 181). This was one of the origins of the so-called chaos theory.

It therefore follows that besides the demonstrative logic that works in certain formal domains, such as in some parts of mathematics, there has to be another kind of logic, of the kind that applies to a decision making process under conditions of uncertainty. Here the final valences will never be true and false, as in the former case, but wise and unwise. Therefore the procedure used to make prudent or wise decisions is not called demonstration but deliberation: there is a demonstrative logic, and another one that is deliberative.

Deliberation is the proper procedure of "practical reasoning". This also goes back to Aristotle (Aristotle 1831, 1139 a 11–30). Economic, political, legal, ethical and technical decisions are always and necessarily of this type: physicians diagnose patients and take decisions under conditions of uncertainty, and that is the reason why the law obliges them to be prudent, rather than not to commit mistakes. The same is true of judges, airline pilots or indeed any other professional.

It has been necessary to reach this point in order to conclude that deliberation is the method of ethics (Gracia 2011a). This is something difficult to assume, because we would all like apodictic moral decisions, demonstrative in nature instead of deliberative, leaving no room for error. But this is not the case. A thousand attempts have been made throughout history to develop an apodictic moral system, and just as many have failed (Gracia 2010).

An interesting question is why there should be this interest in developing an apodictic system or method of ethics. The reason seems entirely psychological. Deliberation is the procedure for making prudent decisions under conditions of uncertainty. Human beings, however, do not like uncertainty, because it creates insecurity and, ultimately, a feeling of unease. The way of protecting ourselves against this, as we said before, is denial. We deny the obvious, seeking absolute knowledge where there it does not exist. Those who do so are incapable of deliberation. Nobody can discuss something when they are full of anguish, or when they are in denial. In order to deliberate it is necessary to assume uncertainty without anguish: only someone who is able to drive a car without distress can lead wisely.

Deliberation appears whenever a human being seeks to take reasonable or prudent decisions on issues of life. The driver of a car has to continually deliberate with himself in order to drive properly, accelerating or braking, turning the wheel to the right or to the left, etc. This type of discussion is completely natural. But when the decision involves a significant risk, especially one that affects other people, Aristotle taught that deliberation should expand its sphere of reference and become collective. There are purely technical group discussions, such as the clinical rounds of any hospital service, and there are legal collective deliberations, as in the courts of appeal, not coincidentally called "tribunals" (in Latin, groups formed by three people). And there are also ethical deliberations. The role of Hospital Ethics Committees (HEC) is to deliberate on conflicts of value in clinical practice, in order to find the most reasonable and prudent solution.

Deliberation entails a complex and difficult learning process. To begin with, a certain degree of knowledge is needed, which will differ according to the subject matter. But that is the easiest part. The human mind is extremely malleable for the acquisition of knowledge until very late in life. It is much more difficult, however, to acquire the right skills, especially because the plasticity of the nervous system for the acquisition of new skills or the modification of old ones is lost very early on. Everybody mispronounces a language if they have begun to study it as an adult, or even as an adolescent. In deliberating certain skills are needed, such as the ability to verbalize one's own views, not only in relation to the facts but, and what is much more difficult, in relation to things that are not entirely rational but which we believe should be reasonable, as in the case of values. Moreover, a person who deliberates must educate their listening skills and be able to accept that he or she is not absolutely right, and that others, saying things that are different or even directly opposed to their views, can be as right as he is, or even more so. Skills, as is well known, can only be trained through practice, and therefore there is no other way of acquiring that of deliberation than by deliberating. And as if all this were not enough, in order to deliberate certain basic attitudes or character traits are also needed, which humans acquire very early on in life and are extremely difficult to modify later on; some personalities are rigid, authoritarian, fanatical, and incapable of deliberating. Simply wanting to deliberate is in itself insufficient: one also has to have the ability to do so.

Deliberation is the process that human beings have to carry out in order to mature their projects and make sound, reasonable, responsible and prudent decisions. And because all human projects include, as we have seen, three aspects, the cognitive, the emotional and the operational or practical, deliberation will also proceed in three steps. First of all we must discuss the facts, reducing uncertainty in this regard, within reasonable limits. We then have to identify the values that are at stake, and the value conflicts that thus arise. And finally, it will be necessary, once a conflict of value has been selected from those that have been identified, to make explicit not only the extreme courses of action but also the intermediate courses, in order to choose from among them the optimal one, which will always be the one that to the greatest extent promotes, or to the least extent harms, the values at stake.

2.9 How, Then, to Proceed?

It follows from the above that a method of decision making is needed, one that significantly differs from that proposed by the theory of rational choice, and also from the methods commonly used in bioethics, such as principlism (Beauchamp and Childress 1979), casuistry (Jonsen and Toulmin 1988), reflective equilibrium (Daniels 1996) or care ethics (Noddings 1984), amongst others. My opinion is that none of these methods do justice to the nature of human projects and the structure of decision-making (Gracia 2001, 2003). All of them are at best unsatisfactory and are perceived as such by professionals, this being the reason why they are rarely used by HECs, and why the latter more often than not limit their analysis to legal questions, avoiding the strictly moral ones. So what should have been an ethical analysis ends up being something else, a legal evaluation of the case, performed by people lacking a professional qualification in law. This loss of specificity of HECs' ethical analysis of the problems is what makes them so weak in the clinical setting, and what also makes the little use made of them by health care professionals wholly understandable. Ethics Committees cannot be understood as a kind of appeal tribunal to avoid prosecution in the event of conflicts, stopping them from finding their way to the courts. If Ethics Committees are ethical, they should have their own specificity. And this can only come from what qualifies them as such, namely ethics. The rest is an exercise in confusion.

We can thus understand why the primary role of ethics committees is not the analysis and resolution of conflicting cases. The most important goal of an Ethics Committee is to promote a new culture and a new way of making decisions, in which not only the clinical facts are taken into account, but also the values at stake, managing them properly. This is a new philosophy, one which will completely change clinical practice. If Ethics Committees are able to promote this new philosophy in their institutions, no doubt they will receive many more cases, perhaps more than they are able to deal with.

To sum up, therefore, the deliberative process of decision-making can be structured as follows (Gracia 2011b):

Deliberation on "facts"

1. Presentation of the case or problem
2. Analysis of the factual data

Deliberation on "values"

3. Identifying ethical problems
4. Choosing the problem to be discussed
5. Identification of the values at stake in this problem

Deliberation on "duties"

6. Identification of possible courses of action

    (a) Identifying the extreme courses
    (b) Identification of the intermediate courses
    (c) Choice of the optimal course

Testing the consistency of the choice

7. Test of legality: is the decision that is going to be taken legal?
8. Test of publicity: would you be able to publicly defend the decision if needed?
9. Test of time: would you take the same decision if you could wait a few hours or days?

Final decision

Is this procedure in itself sufficient to ensure the correctness of a decision? Of course not. No method can guarantee such an outcome. The goal of the procedure is to organize the process of deliberation so that the decisions taken can be considered prudent, wise, responsible and mature, this being the proper aim of the moral life.

References

Aristotle. 1831. *Aristotelis Opera*. Berlin: Reimer.
Beauchamp, Tom, and James Childress. 1979. *Principles of biomedical ethics*. New York: Oxford University Press.
Daniels, Normal. 1996. *Justice and justification. Reflective equilibrium in theory and practice*. Cambridge: Cambridge University Press.
Dickens, Charles. 1854. *Hard times*. The Guttenberg project (http://www.gutenberg.org/files/786/786-h/786-h.htm).
Gracia, Diego. 2001. Moral deliberation: The role of methodologies in clinical ethics. *Medicine, Health Care, and Philosophy* 4(2): 223–232.
Gracia, Diego. 2003. Ethical case deliberation and decision making. *Medicine, Health Care, and Philosophy* 6: 227–233.
Gracia, Diego. 2010. Philosophy: Ancient and contemporary approaches. In *Methods in medical ethics*, 2nd ed, ed. Jeremy Sugarman and Daniel P. Sulmasy, 55–71. Washington, DC: Georgetown University Press.
Gracia, Diego. 2011a. Deliberation and consensus. In *The SAGE handbook of health care ethics*, ed. Ruth Chadwick, Henk ten Have, and Eric M. Meslin, 84–89. London: SAGE Publications.

Gracia, Diego. 2011b. Teoría y práctica de la deliberación moral. In *Bioética: El estado de la cuestión*, ed. Lydia Feito, Diego Gracia, and Miguel Sánchez, 101–154. Madrid: Triacastela.

Jonsen, Albert R., and Stephen E. Toulmin. 1988. *The abuse of casuistry: A history of moral reasoning*. San Francisco: University of California Press.

Kant, Immanuel. 1968. *Grundlegung zur Metaphysik der Sitten. Akademie-Ausgabe Kant Werke IV*. Berlin: Walter de Gruyter.

Lorenz, Edward. 1993. *Essence of Chaos*. Seattle: University of Washington Press. Appendix 1: Predictability: Does the Flap of a Butterfly's Wings in Brazil Set Off a Tornado in Texas?.

Moore, G.E. 1994. *Principia ethica*, Revisedth ed. Cambridge: Cambridge University Press.

Noddings, N. 1984. *Caring: A feminine approach to ethics and moral education*. Berkeley: University of California Press.

Potter, Van Rensselaer. 1971. *Bioethics: Bridge to the future*. Englewood Cliffs: Prentice-Hall.

Chapter 3
A Human Rights Approach to Bioethics

Roberto Andorno

3.1 Introduction

Human rights and bioethics are conceptually and operationally much closer than usually assumed. This is not surprising as both normative frameworks emerged from the same dramatic events: the Second World War, the Holocaust, and the Nuremberg trials. The Universal Declaration of Human Rights (UDHR) of 1948, which would become the cornerstone of the international human rights law, was to a significant extent informed by the horror caused by the revelation that prisoners of concentration camps, including children, were used by Nazi physicians as subjects of brutal experiments. This shocking discovery led the Nuremberg trial to develop in 1947 the famous ten principles for medical research, which have come to be known as the Nuremberg Code. In this regard, it has been reported that "the details revealed daily at Nuremberg gave content to the rights recognized by Articles 4 through 20 of the Declaration" (Baker 2001, p. 242). Similarly, it has been pointed out that "World War II was the crucible in which both human rights and bioethics were forged, and they have been related by blood ever since." (Annas 2005, p. 160) The fact is that at present all major international ethical and policy instruments relating to bioethics adopt a human rights approach.

Taking into account this close relationship between human rights and bioethics, this contribution aims, first, to briefly present the most relevant international human rights instruments dealing with bioethical issues; second, to explore the reasons for this massive recourse to human rights for setting common standards in this field; and finally, to point out the shortcomings of the human rights approach when dealing

---

R. Andorno (✉)
University of Zurich, Zurich, Switzerland
e-mail: roberto.andorno@rwi.uzh.ch

with biotechnological developments such as reproductive cloning and germline interventions, which put at risk, not the life or physical integrity of currently existing individuals, but the integrity and identity of future generations.

3.2 Human Rights Instruments Relating to Bioethics

3.2.1 Core Instruments of International Human Rights Law

Some basic principles that are relevant to biomedical issues can be found in the core instruments of international human rights law. The Universal Declaration of Human Rights (UDHR) of 1948 sets out a list of principles that play a central role in this field: the principle of the "inherent dignity" of "all members of the human family" (Preamble and Article 1); the prohibition of all forms of discrimination (Articles 2 and 7); the right to life (Article 3); the prohibition of cruel, inhuman or degrading treatment (Article 5); the protection of privacy and personal information (Article 12); and the right to health care (Article 25).

The International Covenants on Civil and Political Rights (ICCPR) and on Economic, Social and Cultural Rights (ICESCR) of 1966 develop the rights set out in the UDHR. Among other principles relevant to bioethics, the ICCPR contains one which is fundamental in this field: the requirement of free consent from participants in biomedical research. According to Article 7 "no one shall be subjected without his free consent to medical or scientific experimentation". It should be noted that the fundamental requirement of informed consent, which first rose to prominence with the Nuremberg Code of 1947, responded to the abusive treatment of concentration camps prisoners by Nazi medical doctors. Thereafter, it was included and developed in the Declaration of Helsinki of 1964/2000. However, this latter document does not have a legal nature, as it was issued by the World Medical Association, which is a non-governmental organization. On the contrary, Article 7 of the ICCPR marks the first time that the necessity of free consent for participation in biomedical research is included in an international legally *binding* instrument.

The ICESCR recognizes a right that plays a fundamental part in bioethics: the right to access to health care, which is defined as "the right of everyone to the enjoyment of the highest attainable standard of physical and mental health" (Article 12.1). This same principle can be found in the Preamble of the World Health Organization's Constitution (1946), and in the UN Convention on the Rights of the Child (1989) (Article 24). The right to health care is one of the most important "second generation rights", which are rights of "progressive realization". This means that, by becoming party to the Covenant, a state agrees "to take steps... to the maximum of its available resources" (Article 2.1 of the ICESCR) to achieve the full realization of such rights. Although international instruments do not specify the kind of health care to be provided, the U.N. Committee on Economic, Social and Cultural Rights, the primary body responsible for interpreting the ICESCR, has enumerated the elements of health care services that are essential to the right to health care: availability, accessibility (i.e., provided on a non-discriminatory basis), acceptability (i.e., respectful of ethical and cultural values), and quality (UN Committee on Economic, Social and Cultural Rights 2000).

Several other international human rights instruments include norms that are relevant to bioethics. For instance, the UN Convention on the Rights of the Child (1989) states that "the child shall have…, as far as possible, the right to know and be cared for by his or her parents" (Article 7.1). This provision as well as the principle of the best interest of the child (Article 3) have been often invoked in the debate about assisted reproduction techniques that use donor sperm or egg to support the right of the children conceived through this method to know the identity of their biological father or mother. In this regard, it is interesting to note that, precisely on the ground of the principle set up by Article 7.1 of the Convention, donor anonymity is being gradually abandoned by domestic laws (Blith and Farrand 2004).

3.2.2 The Universal Declaration on Bioethics and Human Rights

The traditional human rights instruments are clearly insufficient to cope with the complex challenges that emerge from biomedical developments and specific common rules are needed in this area. Since health issues and biomedical technologies that accompany them have increasingly a global nature, the response to the new dilemmas should also be global. Aware of the need for minimal common standards, some intergovernmental organizations began in the mid-1990s to promote an international consensus on some basic norms relating to biomedicine.

UNESCO (UN Educational, Scientific and Cultural Organization) has played a leading role in this regard. This is not surprising as UNESCO is at present the only global intergovernmental organization having been involved for decades in standard-setting activity at the intersection of sciences, ethics and human rights. Through the work of its International Bioethics Committee (IBC), this UN agency has elaborated and submitted to its Member States for approval three global instruments relating to bioethics: the Universal Declaration on the Human Genome and Human Rights of 1997; the International Declaration on Human Genetic Data of 2003 and the Universal Declaration on Bioethics and Human Rights of 2005. These three declarations explicitly use a human rights approach to tackle the issues that arise in the biomedical field. Certainly, these three documents are "soft law" instruments which, unlike treaties, are not legally binding for States, at least in the short term. However, it would be a mistake to underestimate their value, not only because, as some studies show, declarations and treaties are in fact complied with to largely the same extent, but also because soft law instruments may in the long term create binding norms, either by leading to a treaty or by being recognized as customary law (Andorno 2013a).

The Universal Declaration on Bioethics and Human Rights of 2005 aims to provide a comprehensive framework of principles that should guide biomedical activities in order to ensure that they are in conformity with international human rights law. The importance of this Declaration lies in the fact that it is the first intergovernmental global instrument that comprehensively addresses the linkage between human rights and bioethics. The instrument in its entirety has been conceived as *an extension of international human rights law into the field of biomedicine* (Andorno 2013a). According to the chairperson of the Declaration's drafting group, the most important achievement of the text consists in having integrated the bioethical analysis into a human rights framework (Kirby 2006, p. 126). As noted by the Explanatory Memorandum to the Declaration, "the Drafting Group also stressed the importance of taking international human rights legislation as the essential framework and starting point for the development of bioethical principles" (UNESCO 2005, n° 11). This document also points out that there are two broad streams at the origin of the norms dealing with bioethics. The first one can be traced to antiquity, in particular to Hippocrates, and is derived from reflections on the practice of medicine. The second one, conceptualized in more recent times, has drawn upon the developing international human rights law. Furthermore, it states: "One of important achievements of the declaration is that it seeks to unite these two streams. It clearly aims to establish the conformity of bioethics with international human rights law" (n° 12).

References to human rights can be found not only in the title itself of the Declaration, but in several of its provisions, especially those stating that respect for human dignity and human rights constitutes one of the aims of this instrument (Article 2.d); that respect for human dignity, human rights and fundamental freedoms embodies the overarching principle of the document (Article 3.1); that the preservation of cultural diversity cannot be invoked as a reason for infringing human rights (Article 12); that Member States should take appropriate measures to implement the Declaration in conformity with international human rights law (Article 22.1); that domestic legislation regarding informed consent, confidentiality of personal data and non-discrimination should be consistent with international human rights law (Articles 6.2, 7, 9 and 11); that limitations on the principles set out in the Declaration, as well as the interpretation of all of its provisions should be in conformity with international human rights law (Articles 27 and 28 respectively).

3.2.3 *The European Convention on Human Rights and Biomedicine*

The Council of Europe too has opted for a human rights approach to develop common biolegal principles. In 1997, this organization opened for signature the Convention on Human Rights and Biomedicine ("Oviedo Convention" or simply "Biomedicine Convention"). The purpose of this instrument is, according to Article 1, "to protect the dignity and identity of all human beings and guarantee to everyone, without discrimination, respect for their integrity and other rights and fundamental freedoms with regard to the application of biology and medicine".

It is worth mentioning here that the Council of Europe was created after the end of the Second World War to promote human rights and democratic values in Europe. This organization was precisely responsible for the elaboration of the European Convention on Human Rights of 1950 and thereafter for the implementation of a series of mechanisms aimed at ensuring the respect for human rights in the Old Continent. It is within this context that the Council of Europe developed the Oviedo Convention, which was opened for signature in Oviedo (Spain) on 4 April 1997. So far it has been signed by 35 Member States of the Council of Europe and ratified by 29 of them, where it has entered into effect. Although the Oviedo Convention is a regional, not a global instrument, its global significance should not be overlooked. It is interesting to note that the Preamble of the Declaration explicitly refers to the Biomedicine Convention. This is worthy of note because it is unusual that UN declarations cite regional instruments as a source. In addition, it should also be mentioned that the Biomedicine Convention has theoretically the potential to extend its applicability beyond European borders, as Article 34 leaves open the possibility of inviting non-member States of the Council of Europe to adhere to the document.

Certainly, the Convention is not entirely original since many of the principles it contains were already included in more general terms in previous international or regional human rights treaties, such as the above mentioned International Covenants of 1966 and the European Convention on Human Rights of 1950 (e.g. the rights to life, to physical integrity and privacy, to access to health care, the prohibition of inhuman or degrading treatment and of any form of discrimination, etc.). Nevertheless, this is the first time that these rights have been developed and assembled in one single binding human rights instrument entirely devoted to *biomedical* issues.

Similarly to the UNESCO Declaration, the Biomedicine Convention includes explicit references to human rights not only in its title, but also in many of its provisions. Article 1 stipulates that the general purpose of the Convention is "to protect the dignity and identity of all human beings and guarantee everyone, without discrimination, respect for their integrity and other rights and fundamental freedoms with regard to the application of biology and medicine." After having emphasized in Article 1 the principle of human dignity, Article 2 gives central priority to the *individual*, whose interest should always prevail over the interest of science and society. Most of the Convention's provisions dealing with specific biomedical issues are indeed conceived within a human rights framework. Among them, the following can be mentioned:

- the Member States' duty to provide "equitable access to health care of appropriate quality" (Article 3);
- the requirement of informed consent for any biomedical intervention, and not only for research (Articles 5–9);
- the value attached to advance directives (Article 9);

- the recognition of the right to confidentiality of personal medical records, and to the right to be informed and not to be informed (right "not to know") about one's health condition (Article 10);
- the protection against genetic discrimination (Articles 11 and 12);
- the rules governing biomedical research on human subjects (Articles 15–18);
- the conditions for organ and tissue donation by living donors for transplantation purposes (Articles 19 and 20).

3.3 Reasons for the Recourse to Human Rights in International Bioethics

As mentioned above, both bioethics and human rights emerged in the aftermath of the Second World War as a response to the same dramatic events. The birth of bioethics can indeed be traced through the history of medical research ethics since the Nuremberg Doctors' Trial of 1947. Similarly, shortly after the end of the War, the international community initiated the efforts to establish what would become the cornerstone of the novel international human rights system –the UDHR– in order to prevent "barbarous acts which have outraged the conscience of mankind" from ever happening again (Preamble of the Declaration). This common origin of bioethics and human rights explain to a large extent the current convergence of both fields in the governance of life sciences and medicine.

In addition to this historical common ground, several other reasons explain the close connection between human rights and bioethics. The first and most obvious one is that, since biomedical activities are directly related to the most basic human rights, such as the right to life and to physical integrity, the right to confidentiality of personal data, the right to non-discrimination, and the right to health care, it is perfectly sound to have recourse to the normative framework of international human rights law to ensure their protection. In spite of all its evident weaknesses and failures, the existing human rights system, with its extensive body of international standards and wide range of mechanisms and international courts, represents a considerable achievement of our time. As a matter of fact, it would be strange that the existing human rights framework could not be used to protect individuals from harm in the biomedical field. As one legal scholar has noted, adopting the language of human rights to govern biomedical issues means moving towards a more comprehensive understanding of the relationships between human health, medicine and socioeconomic and civil and political rights, and public health initiatives (Knowles 2001).

Moreover, the human rights framework facilitates the formulation of universal standards, because international human rights law is based on the assumption that basic human rights transcend cultural diversity. Human rights are conceived as entitlements that people have simply by virtue of their human condition, and regardless of their ethnic origin, sex, age, socio-economic status, health condition, or religion.

In other words, they are held to be universal in the sense that "all people have and should enjoy them, and to be independent in the sense that they exist and are available as standards of justification and criticism whether or not they are recognized and implemented by the legal system or officials of a country." (Nickel 1987, p. 561). In such a sensitive field as bioethics, where diverse socio-cultural, philosophical and religious traditions come into play, the universal nature of the human rights framework is a precious asset.

A common objection to the very idea of human rights applying universally is that they embody a Western liberal-individualistic perspective and are therefore alien to other cultures. Attempting to impose respect for human rights standards on non-Western countries would constitute a form of cultural imperialism. This argument has also been made in the specific field of bioethics, for instance, to criticize the human rights approach followed by the Universal Declaration on Human Rights and Bioethics adopted by UNESCO (Schuklenk and Landman 2005; Schroeder 2005).

Although the philosophical controversy between universalists and relativists is too complex to be adequately covered in this chapter, it can be pointed out that international human rights law has been developed along the last six decades or so by representatives of the most diverse countries and cultures. Therefore, it is hard to claim that it intends to impose *one* cultural standard over others. Rather, it would be closer to reality to say that human rights seek to promote a *set of minimum of standards necessary for human flourishing in our common world*. Furthermore, the universality of human rights is not necessarily in conflict with respect for cultural diversity. Human rights are conceived by international law as flexible enough to be compatible, within certain limits, with respect for the cultural specificities of each society (Andorno 2013b). As a matter of fact, the human rights system allows some local variations, not in the substance, but in the *form* in which particular rights are interpreted and implemented (Donnelly 1989, 109–42).

Another reason for resorting to human rights in this field is that the notion of human dignity, which is the cornerstone of global bioethical norms, is unable alone to provide concrete responses to most challenges raised by biomedical advances. Though respect for human dignity embodies the ultimate ground of biolegal norms (and, in general, of all human rights norms and practices), it is obviously not enough to simply refer to this foundational principle to draw clear answers to most bioethical dilemmas. Rather, some further explanation is usually required. It is necessary to indicate *why* some practices are considered to be in conformity (or not) with human dignity. The need for specification of the dignity principle explains why this concept normally operates through other more concrete notions (informed consent; bodily integrity; non-discrimination; privacy; confidentiality, etc.), which are formulated using the terminology of *rights*.

There is also a practical reason for using a human rights framework to tackle bioethical issues: there are few, if any, mechanisms available other than human rights to function as a global normative foundation in biomedicine (Thomasma 2001), or as a "lingua franca of international relations" (Knowles 2001). In this regard, it has been pointed out that "the human rights framework provides a more useful approach for analysing and responding to modern public health challenges than any framework thus far available within the biomedical tradition" (Mann 1996). The human rights strategy allows "a well-tested and long-established common language, rhetoric and institutional practice to be applied in order to achieve consensus both on the nature of the problem and, ideally, on the form of possible solutions to it" (Ashcroft 2010). While bioethics suffers from the plurality of actors and divergent theories, "human rights offer a strong framework and a common language, which may constitute a starting point for the development of universal bioethical principles" (Boussard 2007). Even insisting on the need to explore alternative normative approaches to bioethics other than human rights, it is acknowledged that human rights are "the strongest vehicle for social change currently available to global bioethics" (Gordijn and Ten Have 2014).

This increasing recourse to a human rights framework to address bioethical issues should however not be understood as meaning that "human rights will subsume bioethics" (Faunce 2005) or render bioethical debates useless. Insofar as bioethics is a part of ethics, it cannot and will never be entirely encapsulated in legal form. Though ethics and law interact in various ways and may significantly overlap with one another, they will always operate as two different normative systems. Legal instruments only attempt to establish a minimal ethics, inasmuch as it is necessary to ensure respect for the most basic human goods. In doing so, the law leaves a broad range of issues open for discussion and to the prudential judgment of the various stakeholders involved in medical practice and research controversies.

3.4 Shortcomings of Human Rights for Dealing with Some Biotechnological Developments

Recognizing the great value of human rights standards for governing biomedical research and practice does not amount to claim that that they can solve all bioethical dilemmas. For instance, the human rights language is clearly insufficient to adequately face the challenge of human reproductive cloning and germline interventions. The claim sometimes made that there is a "right not to be conceived as a genetic copy of another person", or a "right to inherit non-manipulated genetic information" are more rhetorical statements than conceptually consistent arguments. Indeed, rights are claims that belong to *existing* individuals, not to persons who do not exist yet, and even less to humankind as a whole.

Thus, when facing these new challenges, instead of appealing to human rights, it would be more appropriate to argue in terms of the need to preserve the integrity of the human species and our understanding of what it means to be *human* (Annas et al. 2002). In the case of human cloning, what is at stake is nothing less than *biparentality*, that is, the fact that human beings are conceived by the fusion of gametes provided by two different persons, a male ("father") and a female ("mother"). This combination of genetic information from two individuals results in children that differ genetically from their parents and from each other, and are absolutely unique. Offspring resemble their parents, but are not identical to them. This uniqueness of every individual places each of them in a better position to develop his or her own personal identity. In contrast, asexual reproduction produces offspring -clones- that are genetically and physically identical to their parents.

One definite advantage of sexual reproduction is that it contributes to the removal of bad genetic mutations and to put two beneficial mutations together. In addition, it increases the genetic variability in organisms of the same species and, in the long run, allows the best adaptations to the environment to be widespread, especially in changing circumstances. Biparentality is regarded by biologists as a hallmark of evolution and as a key feature of advanced animals. On the contrary, asexual reproduction can be mainly found in plants and in unicellular organisms like amoeba and bacteria. Thus, even from a purely biological perspective and leaving aside any moral considerations, it is hard to see how asexual reproduction could represent a progress for the human species. Rather, it seems well that it would constitute the most dramatic regression that humankind has ever experienced.

Regarding germline interventions, what seems to be at risk is *freedom* from deliberate genetic predetermination by third persons, and, in the long run, the principle of equality between generations. Paradoxically, this freedom closely depends upon the circumstance that each individual's features are more due to *chance* than to *choice*. According to some philosophers, chance in human reproduction can be regarded as a value in itself that needs to be protected against a potential misuse of new technologies (Habermas 2003; Jonas 1985).

International organizations dealing with bioethics are well aware of the weakness of human rights to deal with these potential technological developments. This is why they resort directly to the notion of *human dignity*, which has a broader extension than that of *rights* and may cover the value of humankind as such, including future generations. Three examples illustrate this trend: the Universal Declaration on the Human Genome and Human Rights of 1997, which emphasizes the need to preserve the human genome as a "heritage of humanity" (Article 1), and expressly labels human reproductive cloning and germline interventions as "contrary to human dignity" (Articles 11 and 24 respectively); the UN Declaration on Human Cloning of 2005, which calls on Member States "to prohibit all forms of human cloning inasmuch as they are incompatible with human dignity and the protection of human life" (Paragraph d); the Council of Europe's Convention on Human Rights and Biomedicine of 1997, which prohibits germline interventions on the ground that "they may endanger not only the individual but the species itself," (Explanatory Report to the Convention, Paragraph 89), and the 1998 Additional Protocol to the same Convention, which bans human reproductive cloning on the grounds that it is "contrary to human dignity" (Preamble). It must be noted that human dignity is not used here with its primary meaning, which refers to the inherent value of every human individual, but with a secondary (or derivative) meaning, which relates to the integrity and identity of humankind as a whole. Lacking any other conceptual tool to preserve humankind, intergovernmental organizations appeal directly to human dignity, which is regarded as the last conceptual barrier against the alteration of some basic features of the human species.

3.5 Conclusion

International human rights law has expanded significantly over the past two decades to address the rising number of biomedical related issues. This is patent in the instruments produced by two intergovernmental organizations, UNESCO and the Council of Europe, which explicitly resort to a human rights framework to set up common biolegal principles. Several reasons explain this strategy: the close connection between biomedical issues and basic human rights; the universalistic claim of human rights, which facilitates the formulation of transcultural standards; the fact that the key notions employed at the domestic level to protect people from misuse in the biomedical field are already formulated using the terminology of rights; the lack of any conceptual and institutional instrument other than human rights to produce an international normative framework relating to biomedicine. The emerging international biolaw is not to be regarded as an attempt to subsume medical ethics, or as a strange, and indeed impossible, hybrid between ethics and law. Rather, it should be seen as an extension of international human rights law to the field of biomedicine.

References

Andorno, Roberto. 2013a. International policy and a universal conception of human dignity. In *Human dignity in bioethics: From worldviews to the public square*, ed. Nathan J. Palpant and Stephan Dilley, 127–141. New York: Routledge.

Andorno, Roberto. 2013b. *Principles of international biolaw. Seeking common ground at the intersection of bioethics and human rights*. Brussels: Bruylant.

Annas, George J. 2005. *American bioethics. Crossing human rights and health law boundaries*. New York: Oxford University Press.

Annas, George J., Lori Andrews, and Rosario Isasi. 2002. Protecting the endangered human: Toward an international treaty prohibiting cloning and inheritable alterations. *American Journal of Law and Medicine* 28(2–3): 151–178.

Ashcroft, Richard. 2010. Could human rights supersede bioethics? *Human Rights Law Review* 10(4): 639–660.

Baker, Robert. 2001. Bioethics and human rights: A historical perspective. *Cambridge Quarterly of Healthcare Ethics* 10(3): 241–252.

Blith, Eric, and Abigail Farrand. 2004. Anonymity in donor-assisted conception and the UN convention on the rights of the child. *The International Journal of Children's Rights* 12(2): 89–104.

Boussard, Helène. 2007. The 'Normative Spectrum' of an ethically-inspired legal instrument: The 2005 universal declaration on bioethics and human rights. In *Biotechnologies and international human rights*, ed. Francesco Francioni, 97–127. Oxford: Hart Publishing.

Donnelly, Jack. 1989. *Universal human rights in theory and practice*. Ithaca: Cornell University Press.

Faunce, Thomas. 2005. Will international human rights subsume medical ethics? Intersections in the UNESCO Universal Bioethics Declaration. *Journal of Medical Ethics* 31(3): 173–178.

Gordijn, Bert, and Henk Ten Have. 2014. Future perspectives. In *Handbook of global bioethics*, ed. Henk Ten Have and Bert Gordijn, 829–844. Dordrecht: Springer.

Habermas, Jürgen. 2003. *The future of human nature*. Cambridge: Polity Press.

Jonas, Hans. 1985. *Technik, Medizin und Genetik. Zur Praxis des Prinzips Verantwortung.* Frankfurt: Insel.
Kirby, Michael. 2006. UNESCO and universal principles on bioethics: What's next? In *Proceedings: Twelfth session of the international bioethics committee*, ed. UNESCO. Paris: UNESCO.
Knowles, Lori. 2001. The lingua franca of human rights and the rise of a global bioethics. *Cambridge Quarterly of Healthcare Ethics* 10: 253–263.
Mann, Jonathan. 1996. Health and human rights. Protecting human rights is essential for promoting health. *British Medical Journal* 312: 924–925.
Nickel, James. 1987. *Making sense of human rights: Philosophical reflections on the universal declaration of human rights.* Berkeley: University of California Press.
Schroeder, Doris. 2005. Human rights and their role in global bioethics. *Cambridge Quarterly of Healthcare Ethics* 14(2): 221–234.
Schuklenk, Udo and Landman, Willem. 2005. From the editors. In *Developing world bioethics.* Special issue: Reflections on the UNESCO Draft Declaration on Bioethics and Human Rights 5(3): iii–vi.
Thomasma, David. 2001. Proposing a new agenda: Bioethics and international human rights. *Cambridge Quarterly of Healthcare Ethics* 10: 299–310.
UN Committee on Economic, Social and Cultural Rights. 2000. *General comment no. 14: The right to the highest attainable standard of health (Art. 12).* Geneva: UN.
UNESCO. 2005. *Universal declaration on bioethics and human rights.* Paris: United Nations Educational, Scientific and Cultural Organization.

Chapter 4
Philosophical Imperialism? A Critical View of North American Principlist Bioethics

Carolina Pereira-Sáez

4.1 Introduction

Principlism is an approach to Bioethics developed and clearly expounded by Tom L. Beauchamp and James F. Childress in the book *Principles of Biomedical Ethics*. In this work the authors state that in spite of the current lack of agreement between different ethical perspectives, there are nevertheless four fundamental universal ethical principles that constitute a starting point for bioethical decision-making, even in the most difficult cases (Beauchamp and Childress 2013, 2). In their view these principles are part of the general norms of common morality (Beauchamp and Childress 2013, 12), considered to be the set of norms that all morally serious persons share[1] (Beauchamp and Childress 2013, 3). Beauchamp and Childress consider that the four principles are sufficient for resolving challenging biomedical dilemmas

---

This chapter is one of the results of the research project DER2010-17357, financially supported by the Spanish Ministry for Science and Innovation, and the research project DER2014-52811-P, financially supported by the Spanish Ministry of Economy and Competitiveness.

[1] This criterion is somewhat problematic, and may even be circular: common morality is the set of norms accepted by all morally serious people, i.e. the set of norms accepted by everyone who accepts the norms of common morality. Herissone-Kelly holds that Beauchamp and Childress's understanding of common morality is not descriptive—in which case the issue of circularity would arise—but conceptual (Herissone-Kelly 2003, 65–78; Herissone-Kelly 2011, 584–587). Beauchamp and Childress offer their reply to this issue (Beauchamp and Childress 2013, 416–417, 420 ff).

C. Pereira-Sáez (✉)
Universidade da Coruña, A Coruña, Spain
e-mail: cpereiras@udc.es

and reaching agreement, even between people holding different ethical theories.[2] Principlism thus proposes a biomedical Ethics that is universally applicable, even in the absence of an ethical agreement; in other words, a Bioethics that is not attached to any specific culture, religion, moral theory or metaphysical grounds. Its alleged universal nature is Principlism's most attractive feature (Fisher 2010, 2), and therefore warrants closer scrutiny.

In the absence of any ethical consensus, Principlism thus proposes that practical issues in Bioethics should be resolved by having recourse to common morality, where universally shared general moral beliefs are to be found. However, the four basic principles of common morality to which the authors attribute particular bioethical relevance are in themselves insufficient to resolve the great majority of practical cases. Insofar as they belong to common morality, these principles are abstract and scarce of content, and "abstract norms do not contain enough specific information to provide direct and discerning guidance" (Beauchamp and Childress 2013, 9). The four principles, therefore, although they express rules about what is right or wrong in human behaviour, are only an abstract starting point from which it is possible to develop more specific behavioural norms. Besides that, the need for the principles to be developed before they are applied is due to their equal and only *prima facie* binding character. None of them contains a supreme value that must prevail in every case of conflict between principles; to the contrary, they can all be overridden if in a given case they come into conflict with a norm that creates an equal or stronger obligation (Beauchamp and Childress 2013, 15). In such situations it becomes necessary to locate the greatest balance, the best proportion of right over wrong, by examining the respective weights of the competing *prima facie* obligations. Thus, insofar as they are abstract and *prima facie* norms, these principles have their limitations, which explain "the need to give them additional content" (Beauchamp and Childress 2009, viii; Beauchamp and Childress 2013, 17, 19) by means of specification and balancing guided by the criterion of overall coherence.

North American Principlism is therefore a particular proposal of Bioethics presented as deriving from common morality by means of specification and balancing guided by the quest for the greatest coherence of the set as a whole. Thus, to estimate the contribution made by Principlism, particular attention must be paid firstly, to its understanding of common morality and secondly, to the method to be adopted.

---

[2] Although ethics is by definition practical, the authors refer to their study as an "ethical theory", to convey the knowledge that aims to identify and justify the norms that guide behaviour and allow it to be evaluated (Beauchamp and Childress 2013, 1); they later go on to define other possible meanings of the term (Beauchamp and Childress 2013, 351 ff.).

4.2 Common Morality as Universal Ethics

Although Principlism upholds the universal nature of a set of basic principles and claims theoretical differences to be irrelevant, it is nevertheless true that in practice there is a lack of agreement on serious bioethical issues. Such a lack of agreement, regardless of its roots, is of the utmost practical transcendence in the real life of bioethical practitioners. If we admit the existence of such universal agreement concerning a set of basic principles, we are therefore entitled to ask for an explanation as to the origin of our ethical differences. To put it another way, do not these differences allow us to question the existence of a common morality "applicable to all persons in all places, [so that] we [can] rightly judge all human conduct by its standards" (Beauchamp and Childress 2013, 3), and is authoritative for all communities (Beauchamp and Childress 2013, 4)? Does such a common morality, universally recognised as authoritative, actually exist? Should this not be the case, or if Principlism is not based on such authority, the relevance of its contribution would be greatly diminished: given the prevailing serious ethical disagreements in today's society, the claim to be universally applicable is the explanation for the enormous attraction currently enjoyed by Principlism (Erin 2003, 85–86). In our context of 'hyperpluralism' (Gregory) and general ethical disagreements, the considerable attraction of Principlism is due to its claim to be universally applicable. Beauchamp and Childress would reply that they are not attempting to justify the rightness or authority of common morality, but simply to state that it exists: the existence of a certain number of universally admitted moral norms is, in their opinion, a fact. At the same time, they acknowledge that universal agreement is insufficient to confer authority on norms (Beauchamp and Childress 2013, 418 ff) (thereby opening up a serious issue, which we will refrain from exploring for the moment).[3] Indeed, the authors uphold the view that common morality is simply constituted by moral *beliefs*, not by a set of objective norms or standards that existed prior to such beliefs. Beauchamp and Childress do not attempt, therefore, to put forward reasons to support their claim that common morality is or should be universally accepted, but rather see themselves as describing a de facto situation[4]; the closest they come to

---

[3] If Beauchamp and Childress' reference to common morality is merely descriptive, then "bioethics […] is reduced to the pure description of what people think about different topics, a far cry from the serious systematic study of the reasons that lead us to think of different behaviours as good or bad" (Requena 2008, 21; DeGrazia 2003, 224).

[4] This claim to limiting themselves to describing moral norms is a problematic one. As stated in footnote 1, the criterion of a morally serious person may be considered to be circular. This question is a recurring issue when attempting to make "value-free" practical philosophy. Beauchamp and Childress need some kind of value criterion to determine who morally serious people are. A parallel could be drawn between Hart's view on law and the principlist view on morality: in Hart's opinion, law is ultimately what judges consider to be law; for Principlism, common morality is what morally serious people share. The problem is that in order to know who is a judge we have to know what law is, and in order to know who is a morally serious person we have to know what morality is. On the insufficiency of this value-free perspective for the purposes of practical philosophy, see Pereira-Sáez 2007, *passim*.

offering any kind of justification is their statement that these norms "have proven over time that their observance is essential for stability and civilized interaction" (Beauchamp and Childress 2009, 394).[5] In any event, the authors consider they are merely describing a fact and therefore do not feel obliged to offer any kind of justification of the authoritative nature of common morality. And it is precisely because we are convinced of its authority that "[t]he literature of biomedical ethics virtually never debates the merit or acceptability of these central moral norms" (Beauchamp and Childress 2013, 3). However, even to the authors themselves, seeing the authority of morality as a matter of belief appears insufficient for providing a reasonably satisfactory explanation of common morality; for example, they state that the norms of common morality could change if to do so would benefit human society: a criterion which would demand a judgment on human wellbeing. At the same time the authors, in order to remain faithful to their commitment to "describing beliefs", restrict themselves to noting the absence of any historical example of changes in the fundamental norms of common morality (Beauchamp and Childress 2013, 413).

Let us admit, for the sake of argument, the authority of common morality. However, the existence of a common morality does not imply that the moral norms of every society are the same, but just that the most fundamental and general of the elements of morality are universally shared. And, as the authors state, it is a fact that "debates do occur about their precise meaning, scope, weight and strength […]" (Beauchamp and Childress 2013, 413). In order to be applied, the basic principles of common morality have to be developed, which can be done in many different ways, some of which can even be mutually exclusive (Beauchamp and Childress 2009, 3; Beauchamp and Childress 2013, 19). In Beauchamp and Childress's words, "[t]he reason why the directive in particular moralities often differ is that abstract starting points in the common morality can be coherently specified in more than one way" (Beauchamp and Childress 2009, 16). If common morality, therefore, can be specified in a number of different particular moralities, does the authority of the former then justify the authoritative nature of all its specifications? We will now examine this question in greater detail.

4.3 The Authority of Particular Moralities

Principlism puts itself forward as a determination of the basic principles of common morality with particular relevance to Bioethics, but it is not the only determination, nor do its authors claim it to be the best possible (Beauchamp and Childress 2013,

---

[5] In the latest edition they put forward a different argument: universal agreements about a norm do not justify its authority, but are qualified considered judgments, and thus serve as effective justification, since the justification of a normative theory lies in its overall coherence, which to a great extent depends on its fit with considered judgments (Beauchamp and Childress 2013, 418). In this case the solidity of the justification will depend on our judgment of the method, which will be analysed in a subsequent section of this article.

412). We can ask ourselves, then, whether Principlism is backed by the authority of common morality, and if the same applies to other possible determinations, even those leading to decisions incompatible with those proposed by Beauchamp and Childress.

Beauchamp and Childress's answer to the first of these two questions is not a simple one, and, to my mind, this is because of its transcendence for the second. The authors think that their proposal can be considered as a set of justified ethical beliefs, supported by the authority of common morality (Beauchamp and Childress 2013, 410–411); however, they also state that they do not claim to have the authority of common morality at every level of their proposal (Beauchamp and Childress 2013, 5), nor that their proposal enjoys the universal acceptance that would characterise such a thing as common morality. Principlism therefore lays claim, at least to a certain degree, to the authority of common morality. If, as stated above, we admit that such an authoritative common morality does indeed exist, how then are we to understand that "certain degree"? According to Beauchamp and Childress, the justification of a given specification of common morality depends on the method (Beauchamp and Childress 2013, 17, 19, 404 ff). It would therefore appear that the principlist proposal would be authoritative to the extent to which it *rightly* specifies common morality. Nevertheless, according to the authors, this method is insufficient in itself to justify a particular morality, because coherence –which is the ultimate criterion that guides the principlist method – is not in itself a guarantee of rightness, the necessary starting point being a series of considered judgments that reflect basic norms of common morality, even if they are not definitive.

It would therefore seem that if Principlism is based on a series of right considered judgments and specifies them coherently it would then enjoy the *full* authority of common morality. Consequently, not laying claim to more than a "certain" authority is an option taken by the authors that removes the need to provide a more solid justification without falling into the presumption that the balanced judgements that withstand the method of coherence are the right ones. *Principles of Biomedical Ethics* certainly does not seem to uphold such a claim, nor provide the bases for such an endeavour—as we shall see when we analyze their method.

On the other hand, affirming "a certain degree" of authority allows Principlism to acknowledge other possible legitimate determinations of common morality. This is essential for their approach, since it enables it to avoid falling into "moral imperialism" and put itself forward as being compatible with ethical pluralism[6] (Beauchamp and Childress 2013, 4), whilst simultaneously maintaining its meaning as a bioethical proposal. If it does not claim any authority, if it does not claim to be supported by reasons, which the authors presuppose to spring from common morality, Principlism could not serve as a guideline for the practice of Bioethics (Beauchamp and Childress 2013, 25).

---

[6] According to A. Dawson and E. Garrard, the compatibility between moral "imperialism" with regard to common morality and relativism with regard to private moralities—put forward by Gillon (the European proponent of North American Principlism)—is unsustainable: see Dawson and Garrard (2006, 200–204) and Gillon (2003), *passim*.

From a practical point of view, such a response does not seem to be particularly useful. Common morality needs to be specified, and the authors insist on the idea that "nothing in our method ensures that only one specification or only one line of coherent specification will be justifiable" (Beauchamp and Childress 2013, 411). On the other hand, what common morality offers as a starting point is not only abstract, but also makes it impossible to arrive at a definitive specific norm: Beauchamp and Childress think it a frequent mistake to ignore that exceptions can easily be found to even the firmest specific rules (Beauchamp and Childress 2013, 19). Failing to take into account this non-absolute nature of moral norms is the cause of "many stubbornly imperious pronouncements in biomedical ethics" (Beauchamp and Childress 2009, 19). What we must always bear in mind is that:

> Different parties may emphasize different principles or assign different weights to principles even when they agree on which principles are relevant. Such disagreement may persist among morally committed persons who recognize all the demands that morality makes on them. [...]. We cannot hold persons to a higher practical standard than to make judgments conscientiously in light of the relevant norms and relevant evidence. [...] One person's conscientious assessment of his or her obligations may differ from another's when they confront the same moral problem. Both evaluations may be appropriately grounded in the common morality. (Beauchamp and Childress 2013, 24–25)

It is therefore necessary to acknowledge a legitimate variety of specifications and "we can assess one position as morally preferable to another only if we can show that the position rests on a more coherent set of specifications and interpretations of common morality" (Beauchamp and Childress 2013, 25). These different legitimate specifications stand apart from the norms of particular moralities that violate common morality (Beauchamp and Childress 2013, 5), i.e. from the norms of particular moralities that cannot be understood as being coherent with the norms of the common morality (Beauchamp and Childress 2013, 407). In practice, however, coherence with the norms of common morality is not sufficient for distinguishing between their various legitimate specification and their violations.

This issue sums up the disadvantages we find in the principlist proposal when it comes to explaining the authority of particular moralities. The authors maintain that not all possible bioethical proposals are equally justified—otherwise, Principlism would hardly be Bio*ethics*, the purpose of which is to differentiate between right and wrong behaviour. However, given the abstract and surmountable nature of their principles, it will be hard to find a moral norm that cannot be seen as a specification of one of the principles, the one that triumphs, in a particular case, over the others. Furthermore, to determine whether a particular specification is coherent or not with abstract common morality it has to be specified beforehand, at least to a certain degree, so how can we judge this prior specification's coherence with common morality? Finally, we must remember that, in the principlist method, all considered judgments, including those that reflect the norms of common morality and are taken as a starting point, can at any time be abandoned. Thus, the criterion of coherence with common morality does not seem to be so decisive in justifying certain particular moral norms whilst excluding others.

We should, then, according to Principlism, admit a number of possible legitimate interpretations of common morality, which may even be in contradiction with each other—and hard to distinguish from violations of common morality. We must bear in mind that preferring the most coherent one does not avoid such a possibility; on the contrary: the fact is that different moral principles come into conflict in moral life, and these conflicts at times produce moral dilemmas that cannot be resolved (Beauchamp and Childress 2013, 12).[7] Disagreement as to what the demands of moral norms are, and how they should be applied, is unavoidable (Beauchamp and Childress 2013, 412). Morality is not unitary, the moral life is characterised by conflict and ambiguity, and the method proposed by Beauchamp and Childress does not claim to surmount these features (Beauchamp and Childress 2013, 395–396).

We could therefore think that generally speaking all understandings of Bioethics are legitimate and authoritative, since common morality, as a set of universally shared moral beliefs, is, probably, the starting point for each and every one of them. In other words, we could think that anyone who takes the trouble to put forward a specific understanding of Bioethics is a morally serious person and thus starts from these universally admitted principles. Even what the authors describe as "stubbornly imperious pronouncements" can very probably be understood as an interpretation of the four principles, because their content is extremely limited, as a result of their abstract nature (Beauchamp and Childress 2013, 395). Nevertheless, the authors fail to consider that those who accept particular moralities can enjoy the backing of the authority that common morality provides; they may be legitimate, but are only binding on the members of the group.

It does not appear, however, that Principlism puts forward any definitive reason for not applying this limitation of the authority of common morality to its own understanding of Bioethics. The principlist proposal would thus be saying that each particular morality is binding on the members of the group, which is tautologous to saying that those who accept a particular morality accept it. In my view, if when all is said and done this is what Principlism is proposing, we are justified in coming to the *prima facie* conclusion that its contribution to Bioethics is neither very substantial nor very definitive.

4.4 Two Understanding of Principlism

In order to hold that Principlism makes a significant contribution to Bioethics we have to understand that its proposal claims to be supported by the authority of common morality. However, if the principlist development does not claim to be the best

---

[7] The authors consider that situations might arise in which even the morally preferable action does not enable a person to fulfil all his or her moral obligations and will be deplorable (from the standpoint of moral obligation, not of sentiment, as they take pains to clarify): Beauchamp and Childress (2013, 16).

possible interpretation of common morality, to prioritize it is to indulge in ideological imperialism. So which of these two views of Principlism is the right one?

Principlism's critics reflect both views: in some cases the Four Principles are alleged to be so empty of substantive content that they really fail to work as they are supposed to do. Rather than guidelines that direct actions, they are merely names that evoke matters to be taken into account when dealing with moral issues (Clouser and Gert 1990, 219–223; Clouser 1995, 223–224). In reply, Beauchamp and Childress affirm that the greater the principles' substantive content, the lesser the probability of any agreement on them being reached. What we have, then, is indeed an agreement, but a very superficial one—or, to borrow John Finnis' terminology, a pre-moral one.[8] As Tuija Takala puts it,

> [...] we have the four principles—autonomy, justice, beneficence and nonmaleficence—that we all agree upon. We all think that autonomy is good, that justice is good, that it is good to do good, and that it is good not to inflict harm. In short, by definition, we think that good is good. It is what constitutes the good in various circumstances that we cannot agree upon. (Takala 2001, 73)

Some claim, therefore, that the Four Principles are not much more than moral notions (commonly accepted ones, it is true) that are too vague to serve as a foundation for a universally applied ethics. It is by no means hard to find passages in *Principles of Biomedical Ethics* that support this understanding of Principlism: for example, the affirmation that it is not possible to find any private morality that does not include norms against theft, murder or the breaking of promises; it is only possible to find different understandings of what constitutes theft, what constitutes murder or what constitutes breaking a promise, and which are the exceptional cases (Beauchamp and Childress 2013, 416). This would thus be the first possible understanding of Principlism. It only claims to affirm the authority of common morality, not its principlist specification, hence the critique that Principlism's contribution to Bioethics is a rather meagre one.

However, it is also true that *Principles of Biomedical Ethics* is put forward not as the undisputed affirmation that justice, autonomy, beneficence and nonmaleficence are good things, but as a specification of their demands. Much of the book consists of a specification of the practical demands of the principles (Beauchamp and Childress 2013, chapters 4, 5, 6, 7 and 8). Is this a good enough reason for maintaining that Principlism offers a global, real and feasible Bioethics?

This second understanding of Principlism underpins a second set of critiques, which reject Principlism's very starting point, i.e. the conviction that the Four Principles are universal: "The Georgetown principles do not hold the monopoly of truth in health-care ethics" (Häyry 2003, 208). To my mind, what these authors are attempting to refute is obviously not the universal truth that good is a good thing. Rather, they are expressing their doubts as to whether the understanding of the Four Principles put forward by Principlism is a universal proposal. In other words, they

---

[8] John Finnis states that the first principles of practical reasoning are pre-moral, and the failure to attend to at least one of them is irrational rather than immoral (Finnis et al. 1987, *passim*).

reject the principlist interpretation of good and its practical demands[9] for a variety of reasons, one of these being that they consider it to be a reflection of North American values that would be exercising a kind of "ideological colonisation" over the rest of the world. What these critics reject, then, is that Principlism should be able to arrogate the authority of common morality. Beauchamp and Childress may suggest we put such matters—the substantiation of the principles, the relationship between them, the way of resolving disputes—aside, considering them to be merely a part of ethical theories of little practical interest. However, their own understanding of Bioethics suggests that such matters are indispensable, and that it is almost impossible to take universal agreement over them for granted.

In either of the two most frequent readings of Principlism, therefore, it contributes little to Bioethics—or, to be more precise, it either contributes little or claims to contribute more than it is able to justify. In one of its possible readings Principlism is too abstract, whilst in the other it provides insufficient reasons to take it as authoritative. If what it contributes is the identification of the Four Principles, then it contributes little, since these principles are practically empty of content and can be overridden in specific cases. To affirm that the right action is that which is directed towards good is not saying very much, because an immoral action is also directed in some way towards some kind of good—otherwise it would be irrational rather than immoral. If what it purports to contribute is a definitive bioethical construct with norms and guidelines that can be applied when making decisions, it could then be accused of being an "imperialist morality", because in the final analysis it fails to justify having the backing of common morality to a greater extent than other bioethical proposals.

In order to evaluate Principlism's contribution to Bioethics we must therefore analyse the principlist derivation of specific moral norms from the first principles of practical reasoning—in fact, the entire history of practical philosophy is to a great extent the history of this derivation (Finnis 1983, 69–70). We can state, then, the insufficiency of Principlism, as a result of the disadvantages of the proposed method, which make it impossible to justify how its application leads to the specific practical guidelines the authors put forward. The step from the enumeration of the Four Principles to the content given by the authors when they present specific bioethical criteria is not justified, particularly because their method is based on intuitionism. Thus, the principlist proposal would be attempting to endow a construct based on intuition with "a certain degree" of authority. Only a certain degree, so that a "certain degree" of authority could be attributed also to other intuitions which, precisely because they are intuitions rather than reasons, can lay no solid—i.e., exclusive—claim to authority. To sustain this critique of Principlism, however, we will have to briefly analyse the method it proposes for proceeding from principles to specific decisions.

---

[9] See, amongst others, Takala 2001, *passim*; Holm (1995, 332 ff), Dubose et al. (1994), *passim*; Erin (2003, 85–86), and Campbell (2003), *passim*.

4.5 Specification and Balancing Guided by Coherence

For Beauchamp and Childress the starting point for developing an applicable Bioethics is "to appreciate the limits of the principles" (Beauchamp and Childress 2013, viii); to take into account that none of them is absolute or prevails over the others, that they are merely an insufficient starting point for deciding on specific cases, it being necessary to "give them additional content" (Beauchamp and Childress 2013, viii). In order to give them this additional content, specification and, above all, balancing, are essential: "[t]he problems of bioethics are often problems of getting just the right specification or balance of principles" (Beauchamp and Childress 2013, viii). By means of these techniques Beauchamp and Childress aim to attain coherence or Rawls' "reflective equilibrium" between (to use Rawls' terminology) "considered judgments", amongst which the principles, as well as one's general moral commitments, are to be included, thereby arriving at a "justified moral belief" (Beauchamp and Childress 2013, 381–383). Specification and balancing would thus be guided by the quest for global coherence, starting from a set of considered judgments accepted without any argumentation, drawn from common morality and that could even be abandoned (Beauchamp and Childress 2013, 404 ff). This is the method that determines that certain specifications of common morality are justified whilst others are not.

Principlist specification consists of decreasing the indeterminate nature of abstract norms by reducing their scope of application, establishing by whom, how, when, why, by which means and to whom what is prescribed in the norm should be done. It is therefore a process that is not restricted to analysing the meaning of the norm, but rather one that adds content (Beauchamp and Childress 2013, 17),—essential content, since it even includes the purpose of the action. Norms, however, can be reasonably specified in a variety of ways; in these cases the goal must be the "superior" specification (Beauchamp and Childress 2013, 19), the one that is as coherent as possible with one's own moral judgments and other "relevant moral beliefs". In my view, specification is indeed always necessary (involving, as it does, the exercise of the virtue of prudence), but Beauchamp and Childress do not succeed in fully justifying their understanding of specification, because to do so an objective concept of good is required.

Furthermore, as Beauchamp and Childress state, specification in itself is not enough. Specification refers to the scope or range of each moral norm, but it is often necessary to weigh up the various principles and moral norms against each other. To do so we need to balance the relative weight and force of different moral considerations. This process, according to the authors, consists of looking for reasons that support beliefs about which moral norms should prevail in a given case. So, although specification is usually necessary, balancing is essential, because whilst specification gives rise to a standard that is always applicable to a variety of cases, only balancing makes it possible for all the norms and circumstances affecting a specific complex case to be taken into account. Thus, whilst specification is above all useful for developing demarcated guidelines for action from generally accepted norms, balancing is better suited for judging specific cases (Beauchamp and Childress 2013, 20–22).

The vision of balancing that Beauchamp and Childress put forward has been heavily criticised—their critics include Richardson, who the authors follow on specification—on the grounds that it proposes a way of resolving conflicts that becomes highly subjective (Requena 2008, 22); or because it needs a unit of reference, due to the fact that it involves a comparative weighing up. To my mind, it is the relevance that Principlism gives to the quest for universal coherence based on a set of considered judgments that explains the fact that, to a great extent, the additional content to be given to the principles depends on one's moral intuition. Indeed, the authors consider it an error to claim that the principles can be a thorough and wholly sufficient guideline for action, just as it is an error to ignore the existence of really unresolvable moral dilemmas (Childress 1997, 30–43). Since according to the principlist proposal the principles lack an ethical foundation that makes it possible to go beyond them, our moral judgment must resort to "practical astuteness, discriminating intelligence and sympathetic responsiveness" (Beauchamp and Childress 2009, 22), resources that depend to a large extent on the personal faculties of each moral actor.[10]

Beauchamp and Childress deny that the intuition needed for specification and balancing is just unthinking feeling; indeed, they state that throughout the successive editions of their work they have attempted to reduce the intuitive assignation of balance to the conflicting principles (Childress 1994, 81). They put forward a number of conditions for correct balancing (Beauchamp and Childress 2009, 23), so that infringing one *prima facie* norm in favour of another will only be justifiable if those conditions are respected. However, such conditions appear to be almost tautologous, and are practically restricted to stating that a reason is needed for tilting the balance towards one principle or another (Holm 1995, 336). For Principlism, as emphasised in the principlist reply to the criticisms based on casuistic standpoint, private moral judgements are relatively independent and may lead to the modification or reformulation of general principles (Childress 1994, 87), or at least of the social practices which they provide a shield for.

The method of applying the principles thus runs the risk of cloaking and even "becoming a rhetorical justification of intuitions or prejudices" (Holm 1995, 336), given the serious problems it presents—some of which the authors acknowledge, such as (1) the difficulty of its practical application, (2) the risk of intuitionism, or (3) the problems springing from the search for coherence when morality is seen as non-unitary, so that in the quest for coherence a total reconstruction of the principles of common morality would take place (Beauchamp and Childress 2009, 410, 422–423). There are additional difficulties: all the considered judgments taken as a starting point can be abandoned, thereby making it impossible to overcome the disadvantage that Beauchamp and Childress attribute to coherentism, because their

---

[10] The authors relate this capacity to balance the different moral considerations with moral character, and refer to moral virtues, which they consider to be more in the nature of character traits than habits that can be acquired through conduct (Beauchamp and Childress 2009, 30–63).

claim to possess the "anchor" of a series of substantive principles loses validity.[11] Furthermore, as mentioned previously, in order to assess the coherence of a specification with common morality the content of the latter has to be determined, for which a prior determination is necessary, since its basic norms are abstract and do not include any criteria for resolving conflicts between them (Beauchamp and Childress 2013, 395). The method, therefore, is not only intuitionist but also circular: the starting points are principles that bring together a series of considered judgments that are specified in coherence with the former, seeking the greatest possible coherence amongst the whole. Due to the immense number of mutually incompatible considered judgments that could come into play, coherence is, in fact, impossible, as well as insufficient to justify anything—what is needed is coherence with something that is right and true.

Beauchamp and Childress understand common morality to be a set of commonly shared moral intuitions. The fountainhead of Principlism are not the four principles, but considered judgments, which are pre-theoretical intuitions at varying levels of generality not based on reasons, referring both to specific cases and to more or less general norms. In the authors' opinion they are neither self-evident nor necessarily true, because all of them can be later revised. The strength of these considered judgments lies in the consensus they attract, which Beauchamp and Childress take for granted. They are not real judgments, they are not supported by reasons, because according to Beauchamp and Childress this is in the end impossible: to attempt to justify them on the basis of reasons would necessarily lead us to an infinite regress in their justification. In the final analysis, therefore, the justification Beauchamp and Childress put forward for their norms and case decisions is based on convictions or intuitions. We are thus entitled to ask ourselves how we can explain the binding force of intuitions (based on sentiments rather than reasons), how to proceed from *is* to *ought*. As Requena puts it "if there is nothing else behind the principles, the two tools that we have mentioned will be at the mercy of intuition, carrying the system towards the realms of moral relativism" (Requena 2008, 26).

4.6 Conclusion

Principlism is unable to justify moral norms that could be seen as authoritative determinations of common morality by means of the method it proposes, guided by the search for coherence. It is unable to do so because, as can be seen when this method is analysed, it provides no reasons, basing itself rather on intuitions, in accordance with its premise that the good cannot be rationally known. Yet, if based

---

[11] The ethical code of pirates, for example, may be perfectly coherent (see Beauchamp and Childress 2013, 407). That the coherence of reasoning with a set of substantive principles should lead to the abandoning of these very principles seems somewhat problematic; if so, then the reasoning is not coherent, but this is not enough to deny the justificatory efficacy of the said principles.

on intuitions, Principlism does not work properly, since it is unable to perform the very function of Bio*ethics*, namely to rationally guide action towards the good. However, it would appear that this is not Principlism's intention, but rather to provide a critical reflection on current moral norms and practices (Beauchamp and Childress 2009, 334). In other words, its intention is rather not to help to deliberate on options but to examine the "justifiability" of an option chosen beforehand. Such justification is seen as an acceptance of the option based on intuitions that are to a greater or lesser extent shared, i. e. as an agreement or consensus. However, in the absence of reasons agreement can only be based on force or pressure (along the lines of the 'culture wars' that first appeared in North America and have now spread to Europe). But a proposal supported not by reasons, but by intuitions, is neither communicable nor sharable. Bioethics becomes a matter of conflict solving, of finding solutions based on the consensus reached by the parties concerned.

References

Beauchamp, T., and J. Childress. 2009. *Principles of biomedical ethics*, 6th ed. New York: Oxford University Press.
Beauchamp, T., and J. Childress. 2013. *Principles of biomedical ethics*, 7th ed. New York: Oxford University Press.
Campbell, A. 2003. The virtues (and vices) of the four principles. *Journal of Medical Ethics* 29: 292–296.
Childress, J. 1994. Principles-oriented bioethics. An analysis and assessment from within. In *A matter of principles? Ferment in U.S. bioethics*, ed. E. Dubose, R. Hamel, and L. O'Connell. Peabody: Trinity Press International.
Childress, J. 1997. *Practical reasoning in bioethics*. Blooming/Indianapolis: Indiana University Press.
Clouser, D. 1995. Common morality as an alternative to Principlism. *Kennedy Institute of Ethics Journal* 5(3): 219–236.
Clouser, D., and B. Gert. 1990. A critique of Principlism. *The Journal of Medicine and Philosophy* 15: 219–236.
Dawson, A., and E. Garrard. 2006. In defence of moral imperialism: Four equal and universal *prima facie* principles. *Journal of Medical Ethics* 32: 200–204.
DeGrazia, D. 2003. Common morality, coherence, and the principles of biomedical ethics. *Kennedy Institute of Ethics Journal* 13: 219–230.
Dubose, E., R. Hamel, and L. O'Connell (eds.). 1994. *A matter of principles? Ferment in U.S. bioethics*. Peabody: Trinity Press International.
Erin, C. 2003. Who needs the four principles? In *Scratching the surface of bioethics*, ed. T. Takala and M. Häyry, 79–89. Amsterdam: Rodopi.
Finnis, J. 1983. *Fundamentals of ethics*. Washington, DC: Georgetown University Press.
Finnis, J., J. Boyle, and G. Grisez. 1987. Practical principles, moral truth and ultimate ends. *The American Journal of Jurisprudence* 32: 99–151.
Fisher, A. 2010. Rethinking Principlism: Is bioethics an American plot? *Bioethics Outlook* 21(2): 1–12.
Gillon, R. 2003. First among equals. *Journal of Medical Ethics* 29: 307–312.
Häyry, M. 2003. European values in bioethics: Why, what and how to be used? *Theoretical Medicine and Bioethics* 24: 199–214.

Herissone-Kelly, P. 2003. The principlist approach to bioethics, and its stormy journey overseas. In *Scratching the surface of bioethics*, ed. M. Häyry and T. Takala, 65–78. Amsterdam/New York: Rodopi.

Herissone-Kelly, P. 2011. Determining the common morality's norms in the sixth edition of *principles of biomedical ethics*. *Journal of Medical Ethics* 37: 584–587.

Holm, S. 1995. Not just autonomy. The principles of American biomedical ethics. *Journal of Medical Ethics* 21: 332–338.

Pereira-Sáez, C. 2007. La perspectiva del estudio del derecho en el pensamiento de Finnis: el punto de vista interno desde y frente a Hart. *Archiv für Rechts- und Sozialphilosophie* 106: 130–138.

Requena, P. 2008. Sobre la aplicabilidad del Principialismo Norteamericano. *Cuadernos de Bioética* XIX(1): 11–27.

Takala, T. 2001. What is wrong with global bioethics? On the limitations of the four principles approach. *Cambridge Quarterly of Healthcare Ethics* 10: 72–77.

Chapter 5
Principlism and Normative Systems

Óscar Vergara

5.1 Approach

It is well known that the methodology used to take biomedical decisions known as principlism first originated in the book by Tom L. Beauchamp and James F. Childress, *Principles of Biomedical Ethics*, which has been repeatedly and successfully reprinted since its first edition in 1979 to its latest edition (2013).[1] Currently it is the predominant theoretical model in contemporary Western bioethics[2] and, moreover, even forms part of the tacit premises of Spanish law (García Llerena 2012, XI).

The aim of this chapter is to put forward a series of methodological considerations about this proposal understood as a normative system. Indeed, Beauchamp and Childress formulate four principles which, from this point of view, can be seen as the axiomatic basis of a normative system that enables progressively more specific rules to be derived, in order to provide a method for finding coherent answers to all the different bioethical problems that might arise in the field of medicine. The fundamental issue is whether or not this purpose can be achieved. Therefore, the principlist methodology will be first be examined descriptively, after which its weak points will be highlighted, taking into account subsequent attempts made to strengthen it.

This chapter has been developed within the framework of the Research Project «Principlism and argumentation theory in biomedical decision making» (DER2010-17357), financed by the Spanish Ministry of Education and Science. I wish to thank to Dr. Viviana García Llerena for her interesting commentaries on the long version of this paper.

[1] Hereinafter, unless otherwise stated, all references will refer to this edition.

[2] Even its most tenacious critics recognize its hegemony in the field of biomedical ethics. See Gert et al. (2006, 100).

Ó. Vergara (✉)
Universidade da Coruña, A Coruña, Spain
e-mail: vergara@udc.es

5.2 Description

5.2.1 Elements of the System: An Axiomatic Basis

It is common knowledge that principlist bioethics rests on four cornerstones, these being the famous principles of autonomy, nonmaleficence, beneficence and justice. The idea is that if these principles are systematically deployed, we will be able to find right solutions for ethical conflicts.

(a) The *principle of autonomy* expresses the rule that the individual—the patient, in general terms—must be able to act freely in accordance with a self-chosen plan (Beauchamp and Childress 2013, 101). The basic idea is that *others* will not decide for the patient, whether they are doctors or relatives. Obviously, the patient is not always in a condition to make autonomous decisions, and in this regard a key concept is that of *competence*. And in truth, the principle mandates respect for autonomous decisions. But if the patient is not capable of taking this kind of decisions, his/her representatives will decide for him/her. What is to be the criterion? Sticking to the scope of this principle, Beauchamp and Childress indicate that the right thing to do is to find out what the patient's will is according to the instructions previously given by him/her for the case. If there are no instructions, a substitute judgment: "what would the patient have decided in this case?" could be used. Alternatively, only the criterion of the "best interest" is left (Beauchamp and Childress 2013, 229), which is clearly influenced by beneficence.

(b) The *principle of nonmaleficence* is a classic principle of medicine, long expressed through the aphorism "primum non nocere" (above all, do no harm). This harm may be physical or moral and the rules to specify this principle would be, for example, "do not kill" or "do not cause pain". Harm may be produced fraudulently (intentionally) or unintentionally (by negligence), the latter being understood as the lack of due care (Beauchamp and Childress 2013, 155).

Leaving aside the many clear cases, under this principle we can find a number of borderline cases in which there are doubts as to whether a certain medical treatment is good or bad for the patient, and thus whether to recommend the treatment be withhold or withdrawn. For these cases, various doctrines have taken shape over the years, all of which have been traditionally considered highly plausible. These doctrines are based on the classical distinctions between withholding and withdrawing treatments, between medical treatments and life support treatments and between directly intended effects and those that are merely foreseen. Beauchamp and Childress, however, set aside these distinctions and discuss a more general one, distinguishing between optional and mandatory treatments, according to the quality of life resulting from the balance of burdens and benefits made by the patient or his/her representatives (Beauchamp and Childress 2013, 168–174). This line of thinking leads them, at the end of the chapter, to *conceptually* justify euthanasia (Beauchamp and Childress 2013, 178 ff.).

(c) The *principle of beneficence* is directed at seeking the welfare of the patient. Unlike the principle of nonmaleficence, whose infringement is, *prima facie*, immoral, violating the principle of beneficence is not, according to Beauchamp and Childress, *prima facie* immoral. Nevertheless, in their view this does not mean that it is a mere ideal, since there is beneficent mandatory behavior; for example, helping disabled persons or preventing other people from harm (Beauchamp and Childress 2013, 204).

These authors link this principle in particular to the specific duties associated with the role performed by health care professionals in physician-patient relations. In particular, the awareness of this role in health professionals has secularly justified so-called "medical paternalism". The idea that a health professional is in a better position to determine what will be the best for the patient lies at the heart of this kind of paternalism, which necessarily collides with the principle of autonomy.

In spite of that, not all cases of medical paternalism lack an ethical justification. Beauchamp and Childress distinguish between "soft" paternalism and "hard" paternalism (Beauchamp and Childress 2013, 216 ff.). The former offers fewer difficulties and authorizes the giving of priority to beneficence over patient autonomy in the event of his/her decisions not being voluntary. Hard paternalism is more difficult to justify because it claims the same authorization when the patient decides in informed, voluntary and autonomous terms. Beauchamp and Childress admit they are in line with the principle of soft paternalism, because they understand that it is compatible with respect for the patient's values. Hard paternalism, on the other hand, is on the whole not justified, because it means putting a doctor's values before those of his/her patient. While this may be true, Beauchamp and Childress admit a paternalism of this second kind under certain conditions. A case in point would be that of avoiding a suicide (Beauchamp and Childress 2013, 223–225), another being the one they call "passive paternalism", as in when a doctor does not accept the treatment requested by the patient because it is either considered futile according to the *lex artis* or contrary to his/her conscience (Beauchamp and Childress 2013, 225–226).

(d) Finally, the *principle of justice* refers to the fair distribution of rights and responsibilities, and concerns the kind of justice that we can call "distributive" or "proportional" justice (Aristotle 1999, V. 3). This point has to be made because, strictly speaking, the principles of autonomy and nonmaleficence are really, and clearly, *principles of justice*, although not of the same sort. Under the principle of justice in a restrictive sense, such as Beauchamp and Childress understand it, we must understand fair distribution. This may be performed according to a number of different criteria, depending on which theory of justice is being brought into play; in utilitarianism, for example, the criterion is to maximize utility, whilst liberalism encourages the maximization of freedom and individual property (Beauchamp and Childress 2013, 252–253). This makes it necessary to weigh and specify this principle *according to the individual case*. Beauchamp and Childress do not believe that the standard theories of justice (utilitarianism, libertarian theories, communitarianism or egalitarian theories) allow us to solve this problem for certain, since none of them is wholly valid from all the different points of view, only partially so (Beauchamp and Childress 2013, 262). From this starting point, and within the framework of this principle, they try to face typical problems such as positive discrimination, research with vulnerable individuals or the scope of the right to public health assistance, into which we shall not enter here.

5.2.2 Deploying the System: Specification

The basic content of biomedical ethics is condensed in these four principles. Since they involve the whole field of biomedical ethics, they are too indeterminate in their abstraction in order to be applied with certainty to concrete cases. The demands projected by these principles on a given case can easily be mutually incompatible, thereby creating a conflict. In other words, in the case of medical objection of conscience to the withdrawal of life support measures, as requested by the patient, we may find an opposition between the principle of beneficence or the principle of nonmaleficence (depending on how they are interpreted), which would support the doctor's opinion, and the principle of autonomy, which is intended to ensure that the patient's requests are followed.

However, as this doctrine has a normative nature, it is necessary to reach a certain degree of certainty in order to determine the concrete rules guiding health professionals. Beauchamp and Childress assert that from the principles a series of more concrete "rules" may be extracted, which specify the scope of the former. For instance, a rule extracted from the principle of autonomy is: "respect the autonomy of an incompetent patient following his/her advance directive whenever it is clear and relevant". Furthermore, there are also rules explaining who has the authority to take decisions (i.e. rules for decisions by representation), and rules concerning procedures; for example, to determine suitability for receiving a transplant (Beauchamp and Childress 2013, 14–15).

Due to their essential lack of determination, principles will often conflict, offering the moral agent mutually incompatible courses of action. And for that reason, they have to be shaped for harmonization, and cannot therefore be understood as absolute principles. To develop this concept, Beauchamp and Childress resort to the idea of *prima facie* duties put forward by William D. Ross, who considers that each person has several duties deriving from the different positions he occupies in the relationships carried out with other people (e.g. debtor, husband, friend, citizen, etc.). These duties concur in many cases, presenting different demands that may even be mutually incompatible. Nevertheless, they are *prima facie* duties. Ross points out that when they concur it is necessary to harmonize them, in order to find the definite duty fixing the final pattern of behavior. Thus, we must consider every concrete situation in the most comprehensive way in order to determine which of these duties is the most imperative, given the circumstances (Ross 1988, 19 *et seq.*).

Beauchamp and Childress consider their principles in the same way. They point out *prima facie* demands that must be harmonized in each case. The question that therefore arises is how to perform such a harmonization. Up to the fourth edition of their book (Beauchamp and Childress 1989, 52–54), it seemed to depend, as in Ross, on a sort of intuitive balancing. This gave rise to multiple critiques pointing out the arbitrariness of the criterion (Clouser and Gert 1990, 222–223; DeGrazia 1992, 522) and determined the incorporation in principlism (Beauchamp and Childress 1994, 28–32) of the method of specification proposed by Henry Richardson.

The basic idea of this method is to reinforce the *top down* side of their methodology, introducing a new step between the principles and the cases that apparently provides a certain control of rationality over the decision determining the definite duty. The principles are no longer directly harmonized in the case, but through a series of more concrete rules deriving from them by specification. Since these are rules with a general feature they are therefore applicable to any number of cases, thereby apparently eliminating the problem of arbitrariness and decisionism.

The models of application and balancing, as Richardson explains, assume that moral rules are a fixed and closed group, but this does not take into account that such rules are subject to constant revision. This revision implies the derivation from more general rules of others that are more specific and precisely tailored to the individual cases. The extent of the former is graded by adding clauses determining what, when, why, how, by whom, to whom and with what means (Richardson 1990, 295). Beauchamp and Childress, taking up this idea, affirm that specifying is the operation through which the indetermination of principles is reduced and a more specific content is produced at the time of guiding action (Beauchamp and Childress 2013, 17).

There is a need, however, for a link between the general rule and the specific rule if some type of logical connection between the two is required. The general rule, Richardson affirms, must be recognizable and be reflected in the specific rule. The specification has to contribute to the identity of the moral system. Furthermore, he writes: "*System is more attainable by specification*, because the norms need not be taken as fixed or as formally absolute, and so will conflict less readily and adjust more easily than the sort of norms postulated by a pure model of application" (Richardson 1990, 298).[3]

5.2.3 *(Narrow) Reflective Equilibrium*

However, is specification a rigorous logical operation? Richardson admits that the answer is *no* (Richardson 1990, 301–302), as do Beauchamp and Childress (2013, 24–25). The reason for this is the absence of a single and necessary connection between the general rule and the specified rule. Different specifications——even of a conflicting nature——are perfectly legitimate, particularly in difficult cases.

---

[3] My italics.

Under these circumstances, what degree of control of rationality can specification provide?

The risk of falling once more into intuitionism is apparent. In effect, if specification is naturally open to different possibilities of concretion, the risk that the moral agent decides by intuition and then invokes an *ad hoc* principle for an apparent justification is clear. In order to avoid this, the alternative offered by Beauchamp and Childress is based on Rawls' notion of "reflective equilibrium", which allows a distinction to be made between the different possible specifications. As Richardson puts it, a specification is rationally defensible "so long as it enhances the mutual support among the set of norms found acceptable on reflection" (Richardson 1990, 302). In other words, the person producing a specification should not seek the incorporation of a rule into a system of rules s/he considers valid when the former is incompatible with the latter. Such a rule should obviously not be considered valid unless s/he performs the adjustments needed in the system to avoid the incompatibility (Beauchamp and Childress 2013, 383).[4]

This requirement of coherence within a set of rules is a basic idea in the notion of system. In fact, a normative system that seeks to guide behavior must not include contradictory rules (Alchourrón 1991, 291–301). Two rules are contradictory when compliance with one of them logically excludes compliance with the other (Alchourrón and Bulygin 1977, 297).[5] As Alchourrón and Bulygin assert, there is no possibility of eliminating any contradiction with purely logical resources (Alchourrón and Bulygin 1977, 299). Consequently, if we establish that the most right specific rule is the one that is the most coherent for the group, this implies that we have previously had to accept a set of rules as a valid group. But this acceptance is not logical, something that is also recognized by Beauchamp and Childress when they assert that the selection of the right group of rules may not be done by deduction (Beauchamp and Childress 2013, 393).

This is a problem, because a pirate code of ethics may also hold a coherent group of rules. Nevertheless, we cannot say that their rules are ethically valid. The difference with a valid code of ethics is that the basis of the system should be composed of common morality, this obviously not being the case with a pirate code. However, this control of rationality is limited because, as Beauchamp and Childress admit, common morality may change, although it is not very likely (Beauchamp and Childress 2013, 412–415). This possibility of change radically affects the four basic principles, which are common morality. But, at least within the scope of *narrow reflective equilibrium* (NRE), the eventual modification of the principles would never be the consequence of a theoretical reflection, but would come about as a result of social evolution. Hence, Arras has pointed out that the justification of moral judgments by Beauchamp and Childress is of a hybrid and contradictory nature, being foundationalist with regard to common morality, and coherent with regard to our actual reasoning in practical ethics (Arras 2009b, 18–21).

---

[4] Where the term "permanent revision movement" is used.

[5] Which may happen in the case of mandatory rules, if 'Op' and 'O-p' rules concur, as well as in that of permission rules, if 'Pp' and 'O-p' rules concur.

5.3 Normative Systems. Some Problems in the Principlist Model

As mentioned above, the idea of coherence is a basic idea of the notion of a normative system. For two of the greatest representatives of the theory of normative systems, Carlos E. Alchourrón and Eugenio Bulygin, the distinguishing feature of every normative set is the notion of normative consequence: a normative system is simply a group of premises with normative consequences (Alchourrón and Bulygin 1971, 4). From this viewpoint the principlist system is not only a normative system but also an *axiomatic* system, as its basis consists of a finite number of basic premises, namely the afore-mentioned four principles (Alchourrón and Bulygin 1971, 50).

5.3.1 System Normativity

The principlist system has a "normative" nature, as expressly pointed out by Beauchamp and Childress (2013, 2). But, in fact, this is problematic. Indeed, the four basic principles are formulated from a historic fact: the so-called "common morality". It is *common*, because (presumably) it is not related to a certain specific culture. But it is also *historical*, because it is not the expression of the natural or moral law, understood as an objective source of regulation (Beauchamp and Childress 2013, 4).[6]

The problem is that conceiving common morality as a historical product is the same as affirming that common morality has a purely factual and contingent nature. But the point here is that a fact can never ground a duty. Indeed, a historical common morality may validly stop being *common* and even stop being *morality* at all. But even when this is not the case, the fact of being common does not in itself imply any kind of normativity. Beauchamp and Childress recognize this problem at the end of their work, pointing out that no empirical determination about the existence of universal rules has normative implications (Beauchamp and Childress 2013, 418). In effect, it is a *petitio principii*. From where can we extract the premise "common morality *ought to be*"?

Beauchamp and Childress notice that it is necessary to complete the panorama through an answer of an intrinsically normative nature, resorting to any moral comprehensive theory (for example, utilitarianism, Kantianism, etc.), or through conceptual grounds, resulting from the analysis of the concept of morality (Beauchamp and Childress 2013, 418–422). But they do not try to provide any answer for this purpose, this being a pending matter in their work.

---

[6] According to his judgment, common morality is based on "moral beliefs", and is not based on objective standards existing beyond such beliefs.

With this pending matter, there is no other remedy than to *presuppose* normativity. "Our hypothesis", they write, "is that all persons committed to morality accept the standards found in what we are calling the common morality." (Beauchamp and Childress 2013, 5). The question is why they adhere. We might illustrate this with a clarifying example. The rule "the patient must be told the truth" may find its grounds in the principle providing that the autonomy of the patient must be respected. Which are the grounds, in turn, for this last rule? As common morality, they must rest upon the (presumed) fact that the autonomy of the patient is universally respected. But, is it true, right or valid that the autonomy of the patient *ought* to be respected? Beauchamp and Childress divert from the theoretical point, but accept the practical conclusion. They simply make such an assumption. Consequently, the validity of common morality is a sort of transcendental-logical hypothesis without which the regulatory nature of common morality and all the decisions eventually based on it, fall[7].

The problem does not seem to be faced; rather, it is avoided. As regards the matter discussed above, if principlism does not contain a system capable of providing true answers to problems, we can extract a first approach: its answers, no matter how true they may eventually be, will only reach, from the normative point of view, the category of allegedly valid answers.

5.3.2 Setting the Axiomatic Basis

The construction of an axiomatic system may be carried out in two ways. If the starting point is a deductive system, the problem is to find an axiomatic basis from which it may be deduced. This question will be analyzed in this section. However, if the starting point is a finite group of statements the problem consists of inferring all the consequences of the initial statements or axioms (Alchourrón and Bulygin 1971, 50): this will be analyzed in the following section.

Regarding the first issue, Beauchamp and Childress perform this operation when extracting, from the confused generality of the common morality, the four above-mentioned principles. Is this inference right?

Every normative system must have the following formal features: completeness, coherence and independence (Alchourrón and Bulygin 1971, 61–64).

(a) As regards completeness, normative systems should tend to provide solutions for all the eventual cases. Apparently, it does seem that the four principles are

---

[7] This would be analogous to Kelsen's fundamental norm. For him, Law is not a being but an ought to be. So its validity has to be presumed, since ought to be cannot be derived from being. Therefore, the Constitution is not the supreme norm in the legal order, but the hypothetic fundamental norm (*Grundnorm*) upon which the Constitution bases its validity. This fundamental norm is simply a presupposition or a transcendental-logical hypothesis from which all ruling capacity in Law derives. See Kelsen (1949, 115).

aimed at embracing all the possible cases of biomedical ethics.[8] But a justification is missing. In fact there are alternative formulations. More specifically, so-called "European principlism" provides a different set of principles: autonomy, dignity, integrity and vulnerability. The question is which of the two bases better covers all the possible cases. It may well be American principlism, but it is not justified.

Principlism has been charged with establishing a "local thinking", American Bioethics, "at global level", thereby incurring in "moral imperialism" (Tealdi 2005, 4–5). There may be some truth in this critique, but it is convenient to state that, if this is so, it is not so much due to the formulation of the principles forming the axiomatic base but to their concrete development by Beauchamp and Childress, through the subsequent specifications. This, however, is a problem to be dealt with under the next heading.

(b) As regards coherence, the problem is more serious, since principles are potentially in mutual conflict. This is inevitable, taking into account their nature as principles. Principles have a more indeterminate nature than rules, since, unlike the latter, they do not precisely define an event and the normative consequence. It must be added, as Gert, Culver and Clouser have pointed out, that there is no underlying ethical theory behind the principles that endows them with unity and coherence, but rather each one of them is grounded on a different ethical theory. This is the reason why these authors criticize the fact that the invocation of principles has become a "sophisticated technique for dealing with problems ad hoc" (Gert et al. 2006, 124). The point here is that if the basic principles are potentially in mutual conflict, their harmonization through specification will be given, as we will see, by elements belonging to the *moral background of the agent*.

(c) Lastly, as regards independence or non-redundancy, there are also certain problems. In effect, Beauchamp and Childress have difficulty in delimiting the principles of nonmaleficence and beneficence from each other, in the face of other authors' positions, and therefore devote the first section of the chapter dedicated to nonmaleficence to differentiating it from beneficence. They are thus in opposition to the position adopted by Frankena, who considers nonmaleficence to be part of the principle of beneficence.

Another important aspect to take into account is, as we have indicated, that the principle of nonmaleficence (not to harm another person), as well as the principle of autonomy (allow another person to decide) are, *stricto sensu*, principles of justice, since they refer to the rights of others. Even the principle of beneficence is a principle of justice as regards the prevention of harm.[9]

---

[8] In fact, from a casuistic point of view, Jonsen, Siegler and Winslade establish four parameters for facing ethical problems, which are equivalent to the four principles of Georgetown. See Jonsen et al. (2002), *passim*.

[9] As García Llerena reminds us, the clinical relationship is characterized by *otherness*, since the action (or omission) of the doctor falls on an *alter*, the patient; and, likewise, the decisions of the latter undoubtedly affect the former. It is a justice relationship from which rights and duties arise. See García Llerena (2012, 203–204). In effect, according to Aristotle, "justice, alone of the virtues, is thought to be 'another's good,' because it is related to our neighbour". See Aristotle (1999, V, 1, p. 73).

5.3.3 Drawing Consequences

Once the axiomatic basis has been set, the problem is one of inferring all the consequences from the axioms. According to Alchourrón and Bulygin, the positive norm that gives people over 22 the right to vote is just as normative as its normative consequence that people over 30 have the right to vote. Can the same also be said about specifications?

Richardson mentions that his idea of specification has its roots in Aristotle and Thomas Aquinas (Richardson 2000, 285), although he gives no further indications. The idea of specification may effectively be found in Aquinas when he deals with human law; particularly in the discussion as to whether any law enacted by men derives from the natural law (Aquinas 1947, 1–2, q95, a2). According to Aquinas, the justice of human law, on which its validity depends, is a function of its degree of rationality, the first rule of which is the natural law. Obviously, this entails an ontology, which is absent in the principlist theory, but it is clear that the role performed by natural law, although not equivalent, is analogous to the role performed by the principles in the former. For this reason, the distinction that will be made below is also valid in this case. Following Aquinas, we have said that human law derives from natural law, and we can therefore accordingly interpret that rules derive from principles in an analogous sense.

As explained by Aquinas, this derivation can be done in two ways. (*a*) By *conclusion*, when the contents of human law derive from a judgment or practical syllogism. Aquinas gives an example: the mandate "do not kill" derives from the one requiring "do no harm to anybody". (*b*) By *determination*, when we have an option, a choice between different possibilities that men have to comply with many precepts of natural law. In Aquinas' example, natural law establishes that a person who commits a crime must be punished, but to be punished with in one way or another is a determination of the natural law.

In which of these two ways may rules derive from principles? It has been said above that specification is not a rigorously logical operation, but admits a choice between different possibilities, since this is an operation that amounts to concretizing a set of circumstances that delimit the scope of the principle. This, therefore, is not an example of syllogistic thinking. What we are confronted with is not a type of conclusion, but simply a determination, since there is an option. The difference is very important because it has very important consequences. Thomas Aquinas writes in the same place as referred to above: "Accordingly both modes of derivation are found in the human law. But those things which are derived in the first way, are contained in human law not as emanating there from exclusively, but have some force from the natural law also. But those things which are derived in the second way, *have no other force than that of human law.*"[10]

Applying this distinction to our case, we would obtain that, if specifying means determining, *the validity of each specification will rest on itself*. Specifications can-

---

[10] My italics.

not fall back on the support of the principles, since they are not tautological derivations by way of syllogism. *They will not be part, eventually, of the common morality, being referred to as derivations, but will have to show their own plausibility.*

In order to clarify this issue, it is necessary once again to resort to Richardson, who seems to take the distinction into account when he points out that "subordinate norms may be derived from the initial norms either by deductive subsumption or by less formal causal reasoning" (Richardson 2000, 288). The examples given by him below clearly match those provided by Aquinas, mentioned above. In the first case, he asserts that "do not lie" derives from "do not do anything with intention to deceive". And, for the second, he resorts to an example given by Gert, Culver and Clouser: "do not drive after drinking" derives from "do not kill". For Richardson neither of these ways implies specifying rules, the reason being that in both cases, after the operation, the initial norm remains unchanged, while the specification is a kind of interpretation modifying the content of the norm (Richardson 2000, 289).

He then goes on to establish two conditions for a specification. The first of these is a semantic condition, "extensional narrowing", which requires that everything satisfying the specified norm must also satisfy the initial norm. The second condition, this time a syntactic one, "glossing the determinables", requires that narrowing be done through the incorporation of clauses specifying where, when, why, etc. They are two *interpretative operations* that must be concurrent in order to speak of specification. Richardson considers that narrowing without determining is not specifying, and neither is determining without narrowing.

Through these precautions, Richardson, followed by Beauchamp and Childress, seems to try to avoid what Aquinas pointed out about determination. Thus, a transparent connection between the initial rule and the specification is required. This leads the former to correct the example of specification that the latter had incorporated in their 1994 edition, according to which "follow a patient's advance directive whenever it is clear and relevant" would be a specification of the principle of autonomy. For Richardson, the right specification is: "*respect the autonomy of patients* by following their advance directives whenever they are clear or relevant" (Richardson 2000, 290). This guarantees the above-mentioned connection. Otherwise, the relation between autonomy and prior instructions would be subject to complex justifications: in other words, we would have returned to Thomas Aquinas' version of *determination.*

To sum up, Richardson aims to ensure the validity of specifications by stating that they are not new norms, but the same principles, only this time interpreted. He therefore seeks to prevent specifications from depending on themselves and thereby losing their connection with the common morality.

Beauchamp and Childress require a moral theory to have *output power*, i.e. that it must generate more than a list of previously existing axioms. "A theory has output power," they write, "when it produces judgments that were not in the original database of considered moral judgments on which the theory was constructed" (Beauchamp and Childress 2013, 354). It must be noticed that they do not speak about new rules. And so, we should understand, in line with Richardson, that the system will be composed of a series of principles *and their interpretations*.

This is arguable. Taking the above example, we must ask whether there is actually any substantial difference between the statement "follow a patient's advance directive whenever it is clear and relevant" and "respect the autonomy of patients by following their advance directives whenever they are clear or relevant". The only difference is their linguistic structure. It is true that in the second case we guarantee a certain connection with the original principle, and not with another principle, but this link does not eliminate *the will* to intervene. *There is still an option, will and decision, and therefore a new rule.* The connection may be rational (i.e. if we admit the rationality of the virtue of practical wisdom), but it is not logical.

As a result, we are faced with a case of *determination* as stated by Thomas Aquinas, from which it can be inferred that specifications, as new rules, *will have to justify their validity individually*, without being able to resort to the basic principles of the common morality.

5.4 Towards a Wide Reflective Equilibrium

As was stated in the previous section, the rule "people over 30 have the right to vote" is a normative consequence, a conclusion, of the rule "people over 22 have the right to vote". But the rule "respect the autonomy of patients by following their advance directives whenever they are clear or relevant", as a determination, is a new rule needing a justification. In this case, finding a justification is relatively simple, and agreement will be easy, but this is not always the case, as will become clear below.

For example, the specification "respect the autonomy of patients by following their advance directives whenever they are not contraindicated" may be valid if it is coherent in the context of a moral system A. But the specification "respect the autonomy of patients, following their advance directives, even if they are contraindicated", which is a contradictory rule as regards the previous one, may be perfectly valid in the context of a moral system B. There thus arises a serious problem of incoherence between moral systems, since determinative interpretations of moral principles inevitably revolve around the moral convictions or moral background of the different moral agents.

This is so unavoidable that Beauchamp and Childress have been criticized precisely because, while their theory has changed several times, the same cannot be said of their positions about particular cases (Gert et al. 2006, 101).[11] Their specifications seem to have the seal of their moral particular system and are therefore different from those that a moral agent with a different background would formulate. In this sense, it has been pointed out that principlism and its methodological complements, specification and reflective equilibrium, are only valid between what has been called "moral friends". Thus, it has been said that its merit would lie only in that it provides a heuristics for the dialog between several moral communities (Wildes 1992, 484–485).

---

[11] This being why they say their theory is useful for disguising *ad hoc* decisions.

Beauchamp himself admits that, given its abstraction, the practical content that may be extracted from the principles is very small, with the consequence that this content is subject to *diverging interpretations*. He recognizes that principles are not so much applied as explained and made suitable for specific tasks: "Judgment and decision-making," he writes, "are essential for this interpretative process" (Beauchamp 1994, 959). For this reason, he agrees with Mackie when the latter says that ethics is *invented*, in the sense that moral standards are built over time through agreements and inter-subjective decisions: "What is morally demanded, enforced, and condemned," he continues, "is less a matter of what we discover in already available basic principles and more a matter of *what we decide* by reference to and in development of those principles" (Beauchamp 1994, 960).[12]

In view of the limitations of NRE, the next movement of principlism has leaded to a *Wide* Reflective Equilibrium (WRE). Indeed, a possible way of overcoming the above-mentioned difficulties is to submit the background morality itself to analysis, in such a way that it does not constitute a presumption for taking decisions, but rather another element subject to a critical judgment. This is the idea behind the expectations for a "wide" reflective equilibrium that Beauchamp and Childress incorporate in their methodology from the fourth edition of their book onwards, based on a development that Rawls introduced in the second edition of *A Theory of Justice* (Rawls 1999). It is quite a demanding equilibrium because it connects and contrasts the moral background itself with alternative moral theories (utilitarianism, perfectionism, Kantianism, etc.), theories of procedural justice, theories about the person and even sociological theories of an empirical nature (Arras 2009a, 52).

The point is whether this ambitious methodological program we have just qualified as "quite demanding" is even possible. Arras finds several difficulties in responding positively to this question, but since there is not enough room to develop them herein we will only mention the four difficulties he points out and then add a fifth one.

(a) First, it is not very *practical*. In effect, there are so many requirements needed to reach a wide equilibrium that the task risks "paralyzing our thinking" (Arras 2009a, 49). It must be noted that in order to examine any given problem, it is necessary among other things to analyze and criticize the opinions offered by the standard moral theories, which is no small endeavor, even for a professor of moral philosophy.

(b) Second, the WRE is *too indeterminate*, since no guideline about how to go about this task is offered: how are we to compare opposing theories, adjust them and choose between them (Arras 2009a, 57)? As García Llerena has pointed out, the WRE amounts to a syncretistic integration of other models and forms of bioethical thinking that "does not counteract the criticism leveled against its initial version for lack of systematic character, but only gives it an extended meaning" (García Llerena 2012, 182).

---

[12] My italics.

(c) Third, even if the former difficulties may be overcome, *coherence in itself is not an element that justifies anything at all*, not even in the natural sciences (Arras 2009a, 58).
(d) Fourth, coherentism *dissolves* when it comes into contact with principlism. In fact, the strength of this methodology lies in the four cornerstone principles on which a system of determinations is deployed. However, difficult cases lead the moral agent to connect the case directly with several ethical theories which seek to occupy the space of a particular background morality. But, since there is no preferential element in the reflective equilibrium, adjustment is always reciprocal and multidirectional and even the deepest convictions may be questioned, this means that the principles themselves may be questioned (Arras 2009a, 53).
(e) Finally, it is worth adding an even more radical consideration to this broad panorama drafted by Arras. As is well known, modern hermeneutics has shown that pre-comprehension is a transcendental condition of any comprehension (Gadamer 2010, esp. 270 ff.). So, although somebody can compare their own background moral theory with those of their neighbors, they cannot not avoid *considering the latter in the light of the former*. Seeking to avoid the hermeneutic circle through the use of WRE is as doomed to failure as attempting to step on one's own shadow.

5.5 Conclusions

The purpose of this chapter has been to analyze American principlism from the point of view of normative systems and highlight the difficulties that such an analysis reveals. Notwithstanding the major contribution principlism has made to the field of bioethics, it still presents a number of significant problems. First, it has difficulty in justifying the normative nature of the system because the common morality is characterized in historical terms. Second, the underlying basis of the system presents a number of problems as regards formal features (completeness, coherence and independence). Third, deploying the system through specifications is in fact an operation of determination and not merely one of interpretation. This implies a break in the logical link between the initial rule and the specified one, thereby severing the connection with the common morality. Finally, if NRE has proved to be unsatisfactory for regulating inter-systematic relations due to its relative nature, WRE is not much more satisfactory since, contrary to its claims, it is unable to overcome the hermeneutic circle, among other issues.

References

Alchourrón, Carlos E. 1991. Conflictos de normas y revisión de sistemas normativos. In *Análisis lógico y derecho*, ed. Alchourrón and Bulygin, 291–301. Madrid: CEC.
Alchourrón, Carlos E., and Eugenio Bulygin. 1971. *Normative systems*. Wien: Springer.

Alchourrón, Carlos E, and Eugenio Bulygin. 1977. Incompletezza, contraddittorietà e indeterminatezza degli ordinamenti normativi. In *Logica deontica e semantica*, ed. G. Di Bernardo, 291–307. Bolonia.
Aquinas, Thomas. 1947. *The Summa Theologica*. Trans. Fathers of the English Dominican Province. New York/Boston: Bezinger Bros.
Aristotle. 1999. *Nicomachean Ethics*. Trans. W. D. Ross. Kitchener: Batoche Books.
Arras, John. 2009a. The way we reason now: Reflective equilibrium in bioethics. In Steinbock, B.(ed.) The Oxford handbook of bioethics, 46–71. New York: Oxford University Press.
Arras, John. 2009b. The hedgehog and the Borg: Common morality in bioethics. In Steinbock, B. (ed.) *Theoretical Medicine and Bioethics* 30: 11–30.
Beauchamp, Tom L. 1994. Principles and other emerging paradigms in bioethics. *Indiana Law Journal* 69: 955–971.
Beauchamp, Tom L., and James Childress. 1989. *Principles of biomedical ethics*, 3rd ed. New York: Oxford University Press.
Beauchamp, Tom L., and James Childress. 1994. *Principles of biomedical ethics*, 4th ed. New York: Oxford University Press.
Beauchamp, Tom L., and James Childress. 2013. *Principles of biomedical ethics*, 7th ed. New York: Oxford University Press.
Clouser, K. Danner, and Bernard Gert. 1990. A critique of principlism. *Journal of Medicine and Philosophy* 15: 219–236.
DeGrazia, David. 1992. Moving forward in bioethical theory: Theories, cases and specified principlism. *The Journal of Medicine and Philosophy* 17: 511–539.
Gadamer, Hans-Georg. 2010. *Wahrheit und Methode (1960), Band I*. Tübingen: Mohr.
García Llerena, Viviana. 2012. *De la bioética a la biojurídica: el principialismo y sus alternativas*. Comares: Granada.
Gert, Bernard, Charles M. Culver, and K. Danner Clouser. 2006. *Bioethics: A systematic approach*, 2nd ed. New York: Oxford University Press.
Jonsen, Albert R., Mark Siegler and William Winslade. 2002. Clinical ethics. A practical approach to ethical decisions in clinical medicine.
Kelsen, Hans. 1949. *General theory of law and state*. English edition. Trans. A. Wedberg. Cambridge, MA: Harvard University.
Rawls, John. 1999. *A theory of justice*, 2nd ed. Cambridge, MA: The Belknap Press of Harvard University.
Richardson, Henry S. 1990. Specifying norms as a way to resolve concrete ethical problems. *Philosophy and Public Affairs* 19: 279–307.
Richardson, Henry S. 2000. Specifying, balancing and interpreting bioethical principles. *Journal of Medicine and Philosophy* 25: 285–310.
Ross, William D. 1988. *The right and the good (1930)*. Indianapolis: Hacket.
Tealdi, Juan Carlos. 2005. Los principios de Georgetown: análisis crítico. In *Estatuto epistemológico de la bioética*, ed. V. Garrafa, M. Kottow and A. Saada, 1–6. Unesco: Instituto de Investigaciones Jurídicas. The version available at www.bibliojuridica.org/libros/4/1666/7.pdf. Accessed 2 May 5, 2014.
Wildes, Kevin. 1992. Principles, rules, duties and Babel: Bioethics in the face of postmodernity. *Journal of Medicine and Philosophy* 17: 483–485.

Chapter 6
Types of Action and Criteria for Individualizing Them: The Case of Omission of Life-Saving Care

Pilar Zambrano

6.1 Introduction

Normative systems aim to perform at least two functions when regulating human actions. On the one hand, they intend to guide agents, through intelligible guidelines indicating how to act ex ante, in the field of actions that make up the practice regulated by the system. On the other, they use those same guidelines as criteria by which to judge these same actions a posteriori with a certain degree of objectivity.

Judging an action involves at least three cognitive acts. The first is that of attribution, in which an event is identified as being an action that can be attributed to an agent. This judgment allows such an action to be distinguished from natural and/or involuntary human events: "this death was chosen and provoked by John's doing x". Secondly, we have that of individualization (or identification or specification), which identifies an action as a case of a generic type of action: "John committed murder when he decided to kill x". Thirdly, there is that of liability, by means of which the meritorious or demeritorious nature of the action is identified: "John acted wrongly when he committed murder" (Sánchez-Ostiz 2008, 57).

Guiding action ex ante, as well as judging it a posteriori, are both functional to the final good or purpose of those practices subject to regulation. Therefore, each normative system classifies and judges actions according to a teleological criterion, in view of their impact on the final good, or purpose specific to the practice being regulated.

The identification and discussion of the epistemological grounds of bioethical principles occupy a substantial part of most bioethical handbooks and treaties (Beauchamp and Childress 2013, 3; Childress 1982, preface; Engelhardt 1996, 118;

P. Zambrano (✉)
Department of Private Law and International Relations,
University of Navarra, Pamplona, Spain
e-mail: pzambrano@unav.es

Kass 2002, 2; Kultgen 1995, 7; Sgreccia 2007, 47). Once we realize that both the guiding and judging functions of normative systems become intelligible in the light of the final goods or values inherent to each system, we can understand why the problem of the epistemological status of principles has occupied a large part of, not to say monopolized, the fundamental bioethics debate from its beginnings to the present day.

Nevertheless, the attention paid to principles is clearly non-proportional to the attention received by the problem of classifying actions and the discussion of the cognitive grounds for the individualization or specialization of individual actions (Spaemann 2001, 51). This is surprising when we perceive that anything a theory may gain as regards objectivity in the "upper store" of principles is lost in the "basement" of rules and concrete judgments, when it lacks a strong cognitive theory at this level. It is worthless to argue and proclaim that human life is a good or value recognized on an objective basis and deserving of absolute respect, if subsequently the construction of the meaning of the verb "to kill" is left up to social conventions.

On the basis of this conceptual context, in this chapter we intend to outline and develop the thesis that the intrinsic intelligibility of the types of actions regulated by normative systems is a necessary condition for both the efficiency of their guiding function and the objectivity of their judging function. To support this thesis, we will first analyze the cognitive and semantic assumptions entailed in both the claim that the abstract types of actions are distinguishable, and the claim that individual actions can in fact be individualized or judged. Secondly, these criteria will be applied to the moral distinction between "killing by omission of medical care", and "tolerating an unavoidable death". Afterwards, we will briefly address the criteria for the correct specification or individualization of individual actions and/or omissions, as instances of a specific type of action or omission. Lastly, we will allege that more than just a few of the doctrinal and legal proposals advocating the eradication of the distinction between "orthothanasia" and "passive euthanasia" within the bio-legal context wrongly confuse the perspective of bioethical or moral classification and individualization with that of bio-legal classification and individualization.

6.2 The Intelligibility of Types of Action: Convention or Representation?

The answer to the question regarding the intelligibility of types of action implies an answer to the more generic question of the intelligibility of concepts, which may be restated as follows: Is there anything inherent to conceptual classifications, or are conceptual classifications an expression of social conventions?

Although these issues have been discussed from the outset in the fields of the philosophy of language and that of knowledge, there is a clear split between "realist" and "conventional" semantic theories, in line with the works of Putnam and

Kripke. While the former state that the meaning of concepts is inferred or induced from their reference – i.e. from the nature of the things to which they are applied, conventional theories state that meaning is a purely conventional social construction, with no necessary relation to its field of reference Kripke (1972) and Putnam (1975).

A good way to sum up this split is to be found in Quine's words: "The Aristotelian notion of essence was the forerunner, no doubt, of the modern notion of intension or meaning (…). Things have essences, for Aristotle, but only linguistic forms have meaning. Meaning is what essence becomes when it is divorced from the object of reference and wedded to the world" (Quine 1999, 154). This formula clearly expresses the dilemma dividing traditions: while for a semantic theory of Aristotelian roots the meaning of concepts is necessarily linked to its reference, for a "modern" or conventional theory meaning is a conventional product "wedded to the world" – the reference – that, in itself, lacks any intrinsic intelligibility.

When these considerations are applied to types of action, the question about the origin of conceptual classifications can be reformulated as follows: Are types of actions social constructions divorced from the object of reference, wedded to the world, or do they show a given order, one that to a certain extent is prior to any convention?

If types of action were social constructions from start to finish, the effort of making a conceptual distinction between omissions equivalent to killing and omissions that can be classified differently would not be a true cognitive effort. If there are no intrinsic reasons to distinguish the type "to kill by omission" from the type "to accept an unavoidable death", the cognitive effort would be replaced by a political effort, pushing linguistic dynamics towards the direction each interpreter considers useful.

If, on the other hand, we choose to understand types of action as representations of an intrinsic order inherent to its object of reference – human operations – we then assume that this operative order is intelligible in itself, and the efforts of conceptual distinction are cognitive efforts. It is assumed, in other words, that the criterion for distinguishing between types shows an intrinsically understandable criterion for distinguishing between human operations.

6.3 Teleology and Types of Action

In order to identify this criterion it is necessary to go back a step and recognize that every human action can be analyzed from two perspectives, one interior and the other exterior (Aquinas, Summa Theologiae I-II, q.18 a 6, and q.20).

From an exterior perspective, actions may be analyzed as physical movements that may be described empirically, the analysis of which allows for a new split, depending on whether we consider the biological capacities involved in their deployment, or the succession of causes and effects on the physical world surrounding the action, which, unlike other causal chains, are brought into existence world by an agent (Sanchez Ostiz 2008, 4).

From this dual exterior perspective, man is free to trigger a causal chain or not, but lacks the power to modify it (Vigo 2012, 79). A person who fires a gun triggers a series of causes and effects over his or her own body and the surrounding world, whose development and subsequent deployment are beyond his or her factual scope of interference. On the other hand, the bodily movements involved in actions are distinguished from other physical phenomena through their voluntary origin. We can say that someone "fired a gun", and not that a gun "was fired", when that someone *chose* to initiate a biological and physical causal chain involved in the action of "firing a gun". To use Anscombe's well-known expression, an action x is attributable to agent Y, when Y chose to act "under the description" of x (Anscombe 1963, 7).

Bringing together both approaches to an action, it can be concluded that actions are the physical embodiment – we may call it so – of the choice (an interior one) of a "type of description" of the action.

Now, if *what we do* when we act is *what we choose to do*, the inquiry is necessarily focused on the criteria for distinguishing between the possible objects of choice. Or, what is the same in this context, between physical or exterior descriptions of actions. A first and tempting option is to understand that the physical or exterior description of actions is purely phenomenal, just because it is physical. Thus, the verb "to run" – to use a morally neutral type – would be defined as "a set of physical movements involved in the action of running" and/or as the causation of the effects inherent to the action of running on the agent himself and/or on the surrounding physical world.

The tautology of this solution is evident, as is shown by the new question clearly arising from it: What is the criterion used to unify all the movements and/or physical effects involved in the action of "running" in a single class or type of action? Shouldn't we set this classification aside, and replace it by the conjunctive grammatical conjugation of shorter movements and effects? So, instead of saying "run", we would say, "raise the left foot; then raise it a bit higher; then, start to put it down"; "move the surrounding air at a certain speed"; etc (Nino 1987, 45–56).

This absurd example serves to highlight the following: if the criterion for the physical description of actions were solely and exclusively phenomenal, there are no reasons to cut a chain of physical movements and/or chain of effects involved in each type of action at a certain stage of the chain. On the other hand, when we go beyond this phenomenal world and incorporate it to a trend analysis, we observe that the factor common to all actions is their connection with the purposes of biological capacities – such as local movement in the example of running – and/or with exterior goods affected by bodily movements – such as life, health or honor, in the example of assault.

From a physical or exterior point of view, actions are distinguished or classified, then, according to the biological capacity they deploy and/or the physical goods on which they have an effect. This exterior criterion of description is not, however, independent from the interior perspective: physical types are, rather, the subject matter or object of an interior choice. A physical type described regardless of its condition as the object of interior choice is not in fact a type of action, but, at most, a type of physical phenomenon (Aquinas, ST I-II, q. 18 a. 7 ad. 1; q. 20 a. 3 ad. 6; Keiser 2010, 243).

6.3.1 Physical Classification, Moral Classification and Overlapping Classifications

At least three kinds of conclusions may be reached from the link between types of action and purposes or goods.

The first conclusion is that the dual dimension of the analysis of actions, physical and intentional, exterior and interior, gives rise to a dual criterion of their impact on human goods and, therefore, to a dual basic criterion for their classification.

From a merely physical or exterior point of view, actions are typified according to the purposes or goods to which the biological capacities involved are oriented, and/or according to their impact on other goods that necessarily lie within the scope of the physical influence of the action. When actions are analyzed from an internal perspective, the relevant impact does not affect goods or purposes as such, but rather the moral, and therefore integral, fulfillment of the agent (Aquinas, St, I-II, q.18 a 6 and q.20; Finnis 1983, 33, 1991, 45; Ronheimer 2001, 44–53; Ronheimer and Murphy 2008, 41).

The measure – if we may speak about measures – of this overall impact is conditioned necessarily and exclusively by the agent's choice. In other words, what affects a person's integral fulfillment is not so much the harm done to a good, but *choosing* to do such harm, or more exactly, choosing to perform the description or physical type of action that does the harm in question (Aquinas, ST, II-II q. 64, a 7; Finnis 1998, 141–142, 166; Ronheimer 2001, 395–397)

In this sense, Grisez explains that "choices are related to human goods in two ways. First, in choosing one determines oneself in respect to human goods (see *S.t.*, 1-2, q. 1, a. 3; q. 26, a. 2; q. 28, a. 6; q. 86, a. 1 ad 2). One establishes one's moral identity as open to the good in its fullness or as open only to limited aspects of goodness, not to integral human fulfillment. Second, there is a causal relationship: In choosing, one sets oneself to bring about certain states of affairs (see *S.t.*, 1–2 q. 17, a. 1) (…). In deciding to play golf, for example, one integrates the good of play rightly or wrongly into one's moral self, while at the same time setting oneself to perform the playing of the game" (Grisez 1993, 234–235).

According to this scheme, the exterior or physical classification of the action is a basic classification, on which the interior-moral classification operates, as do other perspectives of classification as ways of assessing the impact of the basic action over any type of good, singular or common. Moral, legal, political, and any other kind of normative types are distinguished from each other by the impact generated by the choice of physical types on each system's own goods (Zambrano 2013, 87).

In the case of morality, the pertinent impact concerns the integral fulfillment of the person.

The second conclusion immediately follows from the above. If types of actions are distinguished according to the good or purpose to which they tend and on which they have an effect, their intelligibility is functional to the intelligibility of these goods. This dependence may be a good explanation for the excess of attention paid to principles in the contemporary bioethical debate. On the other hand, the deficit of intelligibility of the purposes or goods to which they tend and on which actions have an effect is necessarily projected on the intelligibility of types of action. Therefore, a normative theory denying the intelligibility of goods, considered either on an individual and/or comprehensive basis, also denies the intelligibility of the types of actions. And a normative theory assuming a conventional semantic and cognitive conception with relation to the goods or purposes on which actions impact also assumes a conventional conception about the types of action.

The third and final conclusion assumes and extends the two former conclusions: the link of intelligibility between the concepts denoting practical goods and types of action is projected on the conditions of objectivity of the individualization or specification of individual actions. In this order of ideas, it has been correctly pointed out that "One thing should be made clear: if the classification, i.e. the description of the action, is arbitrary, so is the operation of extracting concrete actions from the vital praxis *continuum* (…). From the rejection of universals there necessarily follows the rejection of identifiable individual entities, as P. Geatch, Wiggins, C. Rapp et al. have conclusively proved (…). The aforementioned authors referred to individual things. But their argument is also valid, and even more so, for actions (…). And, naturally, if actions are not identifiable individual units, they cannot be judged in moral terms [nor in legal terms, or from any other point of view]…" (Spaemann 2001, 52).[1]

6.3.2 *Killing by Omission as the Physical Description of the Action and Object of Moral Choice*

From the above considerations we can conclude that the distinction between killing by omission and passively accepting an unavoidable death, as different types of moral actions, rests on the following cognitive and semantic premises:

(a) Moral types of action are types of choices, and choices fall on "physical descriptions of the action" or, what is the same, on physical types of actions.
(b) Physical types are distinguished from each other according to the biological tendencies they deploy, and/or according to the exterior goods on which they have an effect. The intelligibility of basic or physical types of action is, therefore, functional to the intelligibility of the ends towards which the biological

---

[1] The translation is ours.

capacities deployed tend, and to the intelligibility of the goods on which they have an effect.
(c) Moral types are distinguished from each other according to the link between the choice of certain physical types and the integral fulfillment of the person. Their intelligibility is thus functional to the intelligibility of the two opposing conceptual poles that come together in a moral choice: basic physical types and the integral fulfillment of the person.
(d) To the same extent that type intelligibility is functional to the intelligibility of biological tendencies and exterior goods, the objectivity of the specification or individualization of individual moral actions is functional to the intelligibility of types of action.

Assuming this conceptual context, the moral type "to kill by omission" may be tentatively defined as the choice of a physical description of the action consisting in (a) omitting a course of action, (b) whose omission causes one's own death or that of a third person and, therefore, (c) negatively impacts upon one's own integral fulfillment.

This triple division of the moral choice does not describe an instrumental sequence, but rather a relationship of involvement. The physical type or description comprehends, in this sense, every case in which one chooses (i) to kill by suspending or omitting an action, as well as those cases in which one chooses (ii) to suspend or omit a course of action, knowing that such an omission results in death.

The relationship of involvement also applies to the link between the choice of physical type and the choice of harming the agent's moral integrity. In other words, the type does not require the agent to explicitly intend his or her self-denigration, because the choice of killing (by omission) cannot be disassociated from the choice of self-denigration. There are not subsequent purposes to the choice of killing that may change its negative impact over the moral integrity of the agent (Finnis 1991, 55–56; Rhonheimer 2000, 350).

Indeed, there are physical types of action whose choice is contrary to the integral fulfillment of the person under any circumstance, and physical types whose compatibility with the integral fulfillment of the person depends on circumstantial factors. Most choices refer to physical types of the second class: they are courses of action lacking an evident, necessary and invariable link with the integral fulfillment of the person. Thus, the physical type "taking what belongs to another" only integrally affects the person, and becomes the negative moral type "theft", when there is no right to take what belongs to another (Millan Puelles 1994, 115). The choice of a few other physical types, on the other hand, necessarily and invariably affects the integral fulfillment of the person. These include the physical type "to kill" which, of course, does away with any possibility of fulfillment of the subject affected by the action and, for this same reason, invariably impedes, on the part of the agent, the fulfillment of the good of friendship and, more particularly, that of justice.

This necessary and invariable negative link implies that, once an agent realizes that a course of action (including an omission) is a case of the physical type "to kill" but nevertheless chooses it, he also chooses to harm his own integrity. Therefore, there is no possible reason that morally justifies the choice of this course of action. In this sense "absolute moral principles" – which are few in number – are conclusive principles of practical reasoning. Once an agent identifies a course of action as being a case of the physical type "to kill" and, therefore, is forbidden by the corresponding absolute moral principle, there is no need for further moral deliberation. The necessary conclusion is that it is unlawful to choose this description of the action (Finnis 1991, 68, 1998, 164–165; Millan Puelles 1994, 119–120).

6.4 The Meaning of the Physical Type "TO KILL BY OMISSION"

6.4.1 *Efficacy of the Omitted Action*

The moral type "to kill by omission" implies choosing the physical description of the action "to kill through/by one's own failure to act".

The first problem posed by this premise is its self-contradiction: the absence of action is unable to cause anything of a physical nature. Phenomena are, by definition, caused by other phenomena, but not acting is a non-phenomenon or, in other words, the deliberate omission of an act. If this is the case, the moral type "choosing to kill by omission" might be relevant for the few mad or ignorant people who do not understand that nothing can come out of nothing or, to put it another way, that death can never be caused by omission.

In spite of the logical rigor of this reasoning, it nevertheless seems significant to assert, for example, that somebody dies due to voluntarily induced starvation. But, what is significant in this premise?

Although from a strictly biological-causal point of view death by starvation is the consequence of chemical disorders resulting from malnutrition, the point is that there is a significant, although oblique, causal link between starvation – or lack of food – and death. The link is significant because the fact of death is not linked to a mere not-being, but to the absence or lack of a factor present in the normal or ordinary course of events that fulfils a significant function for the continuity of life. It is not the absence in itself that is causally linked with death, but the lack of what *should* be present for life to continue.

This logical-semantic digression is not a mere linguistic game, although it does come close to being one. It is instead a useful path towards finding a first criterion to distinguish which courses of action, among the infinite imaginable courses of action that any agent chooses to omit each time he chooses to act (or not), are connected on a causal basis with death, and therefore constitute the physical type "to kill by omission". In other words, to identify those courses of action providing or maintaining indispensable factors for the performance of a vital function.

According to this, the physical description on which the moral choice "killing by omission" falls might be tentatively defined as the omission or suspension of a course of action implying a decisive factor for the performance of a vital function and which, to this same extent, is effective in halting or preventing the onset of a process that will necessarily end in death.

6.4.2 Feasibility and Legality of the Omitted Action

A good way to assess whether this definition is meaningful is to link it with typical omissions to which it potentially refers, distinguishing them according to the (decreasing) evidence of their causal link to subsequent death:

(a) The omission of ordinary care triggering a previously non-existent lethal process, as in the cases of a hunger strike.
(b) The omission of care that may halt, and definitively revert, a single lethal process, originating independently from the omission, such as not giving a blood transfusion to a patient who – setting aside the eventual need of a transfusion – is in good health.
(c) The omission of care that by itself triggers a new lethal process, which is added to another underlying and slower lethal process, originating independently from the omission. An example would be to suspend artificial nutrition and hydration to a terminal patient who ends up dying due to dehydration and/or malnutrition.
(d) The omission of care capable of halting and reversing a lethal process, originating independently from the omission, and which is added to an irreversible and slower lethal process. A possible example is not to perform resuscitation on a cancer patient suffering from a heart attack, at a time prior to the onset of the lethal process inherent to the disease.
(e) The omission of care that may slow down a lethal process originating independently from the omission, but that cannot reverse it, such as the suspension of chemotherapy to a terminal cancer patient.

Using Hart's terminology, it may be said that case (a) is the "central" or paradigmatic case of causal relation between the omission of care and death – one's own or somebody else's – ; while case (e) is the central or paradigmatic case of a lack or absence of any causal link between cause and result (Hart 1994, 124). Nobody would seriously dispute that death by dehydration and/or malnutrition is linked from a physical-causal point of view with the decision to suspend nutrition and hydration. Nor would it appear to be seriously disputable, either, that when care that is absolutely incapable of reverting a lethal process is omitted the cause of death is the underlying terminal illness and not, on the other hand, the omission of such care. As we move towards the center of the table, however, the causal link becomes more obscure. Thus, even when a lethal process is triggered by an underlying illness, it is not clear that the illness is the only physical cause of death when there are kinds of care capable of reversing the lethal process (case b). Even more arguable is determining what the cause of death is when the omitted care may halt and revert a lethal process, but cannot reverse another underlying and slower process (case d).

If we sharpen our vision, even easy cases become difficult once they are analyzed from the point of view of their moral legitimacy and/or the feasibility of the omitted action. Thus, although death is causally produced in the case of a hunger strike as a consequence of the decision "not to eat or drink", it is not clear that somebody who can force-feed the striker and refuses to do so is an accomplice of this death. Lack of clarity does not affect so much the identification of the biological cause of death – lack of nutrients – as the problem of the moral legitimacy of the omitted course of action – force-feeding-. If we look at the feasibility of the course of action, the same "obscuring" effect is produced. What does *being able* to choose an efficient course of action to save a life mean? *May* we choose an efficient course, when it also implies choosing pain? Or is pain the reason for asserting ensure that *it isn't or it wasn't* possible to choose?

If choosing a course of action is neither possible nor lawful, it does not seem reasonable to claim the existence of a causal connection between its omission and the outcome of death. It may be reasonable to allege a causal relation between the absence of a vital factor and death but not, on the other hand, between the omission of an action and death, because action implies choice, by definition.

These considerations show the mistake of equating causation of death by action and causation of death by omission in a single physical type (Ladd 1979, 164; Rachels 1979, 147). This unification might make sense from a moral point of view, since the choice of killing denigrates the integral fulfillment of the person regardless of the way somebody chooses to kill. But the physical type "killing by omission", however, is significantly distinguishable from the physical type "killing by action".

To sum up, the causal link between an omission and death is strictly predicable from the lack of the necessary factors for the development of vital functions. The causal link between the omission of actions and death, on the other hand, is only predicable from feasible and morally licit actions that effectively contribute a life-supporting factor and halt or prevent a lethal process. The physical description of the type "to kill by omission" might therefore be redefined as the omission of a feasible, licit and effective course of action that is able to halt or prevent a lethal process, by way of providing a life supporting factor.

6.4.3 Feasibility as Proportion

The concept of feasibility of life care embraces a variety of dimensions, which, although not overlapping, condition one another; in particular the technical, economic and physical dimensions. We can point out clear cases of technical and economic accessibility, such as the case of care financed by a public health system (economic feasibility) and commonly applied in public and private health institutions (technical feasibility). On the other hand, there are clear cases of economic lack of accessibility, such as treatments not covered by public health programs or by health insurance policies taken out by the patient or their dependent relatives, the cost of which greatly exceeds their assets, as well as their debt capacity. There are also clear cases of technical lack of accessibility, such as the cases when no health institution, whether public or private, provides the corresponding medical service at the appropriate place and time.

Nevertheless, there is a very wide margin for unclear situations, in which the financial cost of care does not exceed the total assets of the patient or their dependent relatives, but entails expenditures that may have a significant effect on their spending power. Or there are cases in which care is economically accessible, but at the same time present a risk to the continuity of life in the short term, due to its novelty or to the lack of expertise or skills on the part of care providers. This lack of clarity deepens when we attempt to determine the physical cost of medical treatments, given the subjective nature of physical pain. How can we determine in these cases if care or treatment is objectively feasible or achievable? When can we validly conclude that the pain provoked by a treatment makes it non-accessible or non-achievable?

One way of shedding some light on these dark areas is to consider the reference provided by the cases we have defined as being easy, or mainstream. Why do we consider as feasible a treatment or care which only causes fleeting inconveniences, does not jeopardize life and, furthermore, is not expensive? The ground for this judgment is the proportion between the minimum effort its implementation requires, and its effectiveness in ensuring the continuity of human life (Sacred Congregation for the Doctrine of the Faith, 5.5.1980; May et al. 1987, 203).

This proportional relation is not universal, but particular and circumstantial, since it depends on highly variable factors, such as the available medical skills the patient can afford; his/her physical and moral strength, in the circumstances surrounding the decision and taking into account the concrete results – efficacy – that may be expected from the specific care in the case. The contextual nature of the proportion between the cost of care and its efficacy implies invalidating those proposals that rigidly discriminate between types of omissions equivalent to killing and those that are not (Beauchamp and Childress 2013, 173).

Care may be affordable for a patient who has taken out medical insurance, and inaccessible for another who hasn't. The same care may be onerous or affordable for the same person, according to the results expected from its application. It may be onerous if it only pretends to prolong an irreversible lethal process, and highly affordable if it is applied to ensure the indefinite continuity of his/her life.

The same can be said for the remaining variables. The pain and the risk of a treatment are not absolute: it all depends on the results that may reasonably be expected from its implementation, under the concrete circumstances for its application.

In any case, any judgment regarding feasibility does not concern the physical, economical or moral cost arising from the underlying illness, but the costs of the treatments themselves, in relation to the benefit that they are expected to generate for the patient's health (May et al. 1987, 208). More specifically, what is judged is the factual possibility of choosing a course of action, rather than the quality of life such a course of action might sustain.

Returning to the definition of the physical type on which the moral choice "to kill by omission" falls, the analysis of feasibility gives rise to a third assertion: what is chosen to be omitted is a course of action that is (a) efficient to halt or prevent a lethal process, (b) technically feasible and (c) accessible for the agent at a cost (physical, moral, economical) that is proportional to its efficiency.

Choosing to omit or suspend care that fulfils these features entails choosing death resulting from the failure to act, even – or all the more so – when the omission is based on a value judgment regarding the (low) quality of life that such care (effective, feasible and proportionate) makes it possible to sustain. On the other hand, when we choose to omit or suspend a course of action due to its lack of efficacy in continuing or maintaining life, or because its costs are judged to be disproportionate to its efficacy, death is not chosen in itself. This choice entails, instead, accepting the inevitable limitations of human power and the finite nature of life.

6.5 Moral Legitimacy of the Omitted Action

Killing by omission is choosing to omit an action combining three physical qualities: such an action must (a) be effective in halting or preventing a lethal process, (b) be technically feasible and (c) come at a physical and economical cost that is proportionate to its effectiveness.

Condition (a) fulfils the conceptual function of including the omission within the kind or physical type of the action of killing. Conditions (b) and (c), however, fulfill the conceptual function of locating the omission within the scope of human actions, as opposed to physical phenomena. Notwithstanding the efficacy of the imaginable courses of action in preventing or halting a lethal process, their omission may only be deemed an action when they are feasible courses of action and, therefore, can be chosen. Conditions (b) and (c) determine, in short, the factual conditions of the possibility of a course of action, and together with condition (a) are necessary and sufficient elements to define the physical type whose omission is equivalent to "killing".

Nevertheless, an action may be eligible in factual terms but forbidden by moral or religious norms, examples being blood transfusion in certain religious systems or the use of drugs which are feasible and effective for the cure of terminal illnesses, but whose production involves the processing and discarding of human embryos.

In order for a course of action to be truly "eligible", it is not enough simply to meet the factual conditions of feasibility, but it is also necessary to add its normative, moral or religious nature. A person who considers that a course of action is forbidden by moral or religious rules does not choose death, even when it is an effective and feasible course of action whose cost is proportionate to its effectiveness.

6.6 From Type Intelligibility to Individualization Fallibility

Returning to the reflections that opened this chapter, morality's capacity to guide as well as to judge depends on the ability of moral agents to comprehend the meaning of moral types of action. In the issue we are dealing with here, whether or not bioethics is able to guide agents' actions in the field of life care depends on the meaning of the type "to kill by omission" having a hard core of intelligibility that cannot be reconstructed indefinitely.

In order to identify this hard core, we have analyzed the physical description of the object of choice of the moral type "to kill by omission", at successive but also complementary levels of precision. The result thus obtained can be summed up in the following definition: killing by omission consists of omitting or suspending a morally legitimate course of action that is effective in halting or preventing a lethal process, at a cost proportionate to its effectiveness.

This description is intelligible only to those who accept the intrinsic intelligibility of the good "life" and the concepts "causation", and "effectiveness". If the concept of life were non-intelligible, the opposing concepts of death and lethal process would not be understandable, either. And, if we were unable to understand the concept of causation and its derivative, effectiveness, the criterion regarding the effect of the physical movements that are omitted on the good or *telos* of life would vanish.

In this sense, it is not at all surprising that the struggle to legalize euthanasia in any of its forms is advocated based on a radically conventional semantic theory, regarding both the concept and the value of human life and its conceptual derivative, the physical type "killing". What is claimed from this perspective is not so much the legalization of "killing", but the freedom of each agent to decide, according to a personal axiological scale, the meaning and the reference of the good "life", and the corresponding meaning of the verb "to kill" (Dworkin 1994, 27–28).

It is obvious that the cost of this sort of privatization of moral language is its intelligibility and the corresponding loss of objective grounds for the individualization of concrete actions. If the concepts of life and "killing" are arbitrarily constructed by each person according to his or her personal criteria, with cognitive basis whatsoever, all possibilities of objectively identifying singular actions as instances of the type "to kill" disappear (Spaemann 2001, 53). And the even higher cost of semantic privatization is quite simply the loss of the guiding and judgmental power of any normative system.

Be that as it may, the guiding capacity of systems is not in itself guaranteed by the intelligibility of types. In fact, there is no point in the agent grasping the meaning of types, if he or she is incapable of recognizing under which circumstances individual actions are adequate instances of them. Guiding the action not only entails the agent knowing the meaning of the moral type "to kill by omission", but also that he or she has moved from the conceptual stage to the judgment stage, for all the relevant aspects of the type.

So, no matter how evident the imperative nature of the rule "not to kill", nor how evident the meaning of the type "to kill", this evidence does not extend to the prudential circumstantial judgment individualizing individual actions (or omissions) as instances of the type. And this is because any movement from the cognitive conceptual stage to the judgment stage involves what Thomas Aquinas calls «the right estimate» of the particular situation, which, insofar as it has to do with particular and variable issues, is highly fallible (Aquinas, ST I-II q. 14 a.3; II-II q. 47 aa. 3, 9; Vigo 2012, 81).

The probability of the fallibility of the individualization of an action is even greater within the biomedical field, where all three sub-judgments of effectiveness, feasibility and proportion constitute applications of extremely complex scientific and technical concepts, in widely varying specific situations. The undeniable fallibility of individualization does not however run contrary to the objectivity or semantic realism of the type, but rather reinforces it. If there is any room for error it is precisely because, previously, it is possible to distinguish between what is right and what is wrong.

References

Anscombe, G.E.M. 1963. *Intention*, 2nd ed. Cambridge, MA: Harvard University Press.
Aquinas, Thomas. *Suma Theologie* I-II: qq. 17–20; 18, a. 7 ad. 1; 20 a. 3 ad. 6; II-II: qq. 49 a. 2, ad. 1; 64, a. 7.
Beauchamp, Tom L., and James F. Childress. 2013. *Principles of biomedical ethics*, 7th ed. New York: Oxford University Press.
Childress, James F. 1982. *Who should decide? Paternalism in health care*. New York: Oxford University Press.
Dworkin, Ronald. 1994. *Life's dominion. An argument about abortion, euthanasia and individual freedom*. New York: Vintage.
Engelhardt, J.R. 1996. *The foundations of bioethics*, 2nd ed. New York: Oxford University Press.
Finnis, John. 1983. *Fundamentals of ethics*. Oxford: Clarendon Press.
Finnis, John. 1991. *Moral absolutes*. Washington, DC: The Catholic University of America Press.
Finnis, John. 1998. *Aquinas, moral, political, and legal theory*. Oxford: Oxford University Press.
Grisez, German. 1993. *The way of the Lord Jesus. Living a Christian life*. Chicago: Saint Paul's Alba House.
Hart Herbert, L. 1994. *The concept of law*. Oxford: Clarendon Press.
Kass, Leon R. 2002. *Life, liberty and the defense of dignity*. San Francisco: Encounter Books.
Keiser, K. 2010. The moral act in Saint Thomas. A fresh look. *The Thomist* 74: 243.
Kripke, Saul. 1972. Naming and necessity. In *Semantics of natural language,* ed. David Davidson and Gilbert Harman, 253. Dordrecht: D. Reidel.
Kultgen, Johh. 1995. *Autonomy & intervention. Parentalism in the caring of life*. New York: Oxford University Press.
Ladd, J. 1979. Positive and negative euthanasia. In *Ethical issues relating life & death*, ed. J. Ladd. New York: Oxford University Press.
May, William, Robert Barry, et al. 1987. Feeding and hydrating the permanently unconscious and other vulnerable persons. *Issues in Law and Medicine* 3: 203.
Millan Puelles, Antonio. 1994. *Ética y Realismo*. Madrid: Rialp.
Nino, Carlos S. 1987. *Introducción a la filosofía de la acción*. Buenos Aires: Eudeba.

Putnam, Hilary. 1975. The meaning of meaning. In *Mind, language and reality*, ed. K. Gunderson, 131. Cambridge: Cambridge University Press.

Quine, W.V.O. 1999. Two Dogmas on empiricism. In *Concepts*, ed. E. Margolis and S. Laurence, 154. Cambridge, MA: MIT Press.

Rachels, J. 1979. Euthanasia, killing and letting die. In *Ethical issues relating life & death*, ed. J. Ladd. New York: Oxford University Press.

Rhonheimer, Martin, and William F. Murphy (eds.). 2008. *Perspective of the acting person: Essays in the renewal of thomistic moral philosophy*. Washington, DC: Catholic University of America Press.

Ronheimer, Martin. 2001. *Perspektive der Moral. Philosophische Grundlagen der Tugendethik*. Berlin: Akademie Verlag.

Sánchez-Ostiz, Pablo. 2008. *Imputación y teoría del delito. La doctrina kantiana de la imputación y su recepción en el pensamiento jurídico-penal contemporáneo*. Buenos Aires: IB de f.

Sgreccia, Elio. 2007. *Manuale di Bioetica. Volume I Fondamenti ed ética biomedica*, Quartath ed. Milano: Vita e pensiero.

Spaemann, R. 2001. *Grenzen. Zur etischen Dimension des Handlens*. Suttgart: Klett-Cotta.

Statement *"Iura et bona"* on euthanasia by Sagrada Congragación para la Doctrina de la Fe, (5.5.1980).

Vigo, Alejandro. 2012. Deliberación y decisión según Aristóteles. *Tópicos* 43: 79.

Zambrano, Pilar. 2013. Principios fundamentales como determinación de los principios morales de justicia. Una aproximación desde la distinción entre la perspectiva moral y la perspectiva jurídica de especificación de la acción humana. In *Ley, moral y razón. Estudios sobre el pensamiento de John Finnis a propósito de la segunda edición de Ley natural y derechos naturales*, ed. Juan B. Etcheverry. Mexico: UNAM.

Chapter 7
Bioethics, Deliberation and Argumentation

José-Antonio Seoane

Health care is not only a question of clinical facts, but also involves certain axiological and normative elements. Due to its ability to integrate these different elements in the clinical decision-making process, deliberation is considered to be the method of medical ethics (Gracia 2001a) leading to the development of a deliberative procedure (Gracia 2001a, b, 2003, 2004; Gracia and Rodríguez Sendín 2006) that has been widely adopted by bioethics committees.

After first characterising deliberation and explaining its relation to bioethics, I look at the structure and distinctive features of the deliberative method. I follow this with a critical assessment of the method, and finally present a revised version containing additional elements derived from hermeneutics and from the theories of legal argumentation.

7.1 Deliberation in Bioethics

The deliberative procedure used by bioethics committees is characterised by a series of features that explain the reason for its existence, its structure and its functioning.

This chapter is one of the results of the research project DER2010-17357, financially supported by the Spanish Ministry for Science and Innovation, and the research project DER2014-52811-P, financially supported by the Spanish Ministry of Economy and Competitiveness.

J.-A. Seoane (✉)
Research Group Philosophy, Constitution and Rationality, School of Law,
Universidade da Coruña, A Coruña, Spain
e-mail: jose.antonio.seoane@udc.es

1. Biomedical decision-making has to be based on a method, since ethical issues cannot simply be ignored or dealt with intuitively, and neither clinical experience, the dictates of conscience, common sense or even imitation are in themselves sufficient. What is needed is a process for identifying, analysing and deliberating on the facts, values, rights and norms involved in healthcare practice, one that provides a rational justification for the decisions.
2. Deliberation is a method, i.e. a way, to reach a decision (*metá*, towards; *hodós*, road). The deliberative method provides a stable and systematic criterion for rational decision-making, free from the vagaries of chance and *ad hoc* choices. It is, however, purely instrumental for achieving a greater end, this being a prudent and wise decision that can provide a solution for a real case.
3. The deliberative procedure is more than just a method for biomedical decision-making; it is an instance of the conception of bioethics as civil ethics (Gracia 2001c, 230, 2005). Deliberation is the logic of daily life (Gracia 2001b, 30–31), a pedagogical tool for each person's life and for building our society (Gracia 2011b, 108–113) that lies at the heart of ethics, politics and practical philosophy (Gracia 2001b, 30–32).
4. Deliberation is the mode of knowledge that characterises practical reason, in which there is no room for apodictic or demonstrative knowledge, only probable knowledge (Aristotle 1999). The method is deliberative since it is a form of practical reason in the rhetorical sense, as a deliberative genre (Aristotle 1926, 1358b7–9) that seeks to recommend the best alternative by combining rhetoric and dialectics (Gracia 2001b, 42), and in the ethical sense as the practice of prudence and balancing the conditions surrounding a decision before acting (Aristotle 1999, 1140a25–b5) (Gracia 2001b).
5. There are no *a priori* solutions, nor a science or *techné* of practical reasoning for clinical cases (Aristotle 1999, 1104a3–8). The deliberative method is prudential: it pertains to the sphere of dialectics and requires the exercise of prudence, understood as both an intellectual virtue and practical wisdom (Aristotle 1999, 1140a23–1140b34). Between the certainty of apodictic knowledge and mere opinion, prudent deliberation reconciles what is general with what is unique, seeking to establish the fair middle point (Aubenque 1963, 64). It is a creative process for knowledge (Gracia 2001b, 42) that flexibly adjusts the general framework to fit the peculiarities of each new case, which always differs in one way or another from all previous cases (Gracia 2011a, 86).
6. Although the final decision is made after a conflict of values has been identified, the deliberative method itself seeks to establish harmony rather than conflict. Conflict is only irresolvable when radical approaches are involved, whilst the aim of prudent and wise decisions is to apply both positive values to the greatest possible extent at the same time.

Conflict lies in reality, not in the method. The healthcare professional (Level I) is experiencing a conflict that the committee expresses using the language of values (Level II), before deliberating and dissolving this conflict of values and duties with the best possible solution (optimization: level III) for the initially conflicting values.

7. The deliberative method is problematic rather than dilemmatic. It does not simply identify two desirable options, since reducing deliberation to extreme courses of action is a falsification of reality, which is always more complex and displays a variety of intermediate possibilities (Gracia 2011a, 92–93). However, although extreme courses of action are not desirable, since they prevent the implementation of one positive value and the ethically optimal decision, identifying them does provide us with the ethical boundaries of the decision.

    The problematic nature of the method separates it from dogmatism and the thesis of the one right answer, both from a logical perspective, as a result of its connection with dialogue and dialectical reasoning, and from that of the anthropological unsuitability of the dilemma, which involves an erroneous simplification of reality (Gracia 2001c, 229).

8. Deliberation is a complex activity that consists of a method of reasoning about facts, values and duties that takes abstract principles, specific circumstances and foreseeable consequences into consideration. It does not merely concern itself with reasons, but also with feelings, values, beliefs, traditions and expectations, all of which form part of our moral decision-making as human beings (Gracia 2010, 68), and does so by using axiological and ethical language, not commonly found in clinical practice.

We have not been trained for deliberation, which is a form of reasoning based on probability or plausibility requiring certain habits and qualities: intellectual humility, the ability to listen and the exercise of prudence (Aristotle 1999, 1003a–1004a; Gracia 2011a, 88, 93). The attitude and habit of deliberation demand our respect for the other person as a valid interlocutor and distance us from fanaticism (Gracia 2011b, 124–129), confirming both the need for and value of other perspectives, which reveal varying approaches to reality (Gracia 2001c, 229, 2007, 8).

7.2 The Deliberative Method: A Description

The deliberative method has brought about a change in the language of bioethics, in that the original language of ethics was neither one of principles nor of rights, but rather one of values, one that is richer and more complex, flexible and even ecological (Gracia 2007, 7–8). This leads us to consider that deliberation is not only a method for ethics or bioethics, but for practical reasoning in general, which is tantamount to saying for human reasoning as a whole (Gracia 2007, 8, 2011b, 120).

The deliberative method rejects theoretical fundamentalism and the pragmatism of decisionism (Gracia 2007, 6). Its adoption as a procedure means abandoning the theory of the four bioethical principles (Beauchamp and Childress 2013), which is a simplification of the wealth and variety of moral reality. Decision-making is not a mechanical task that consists of establishing a hierarchy of principles and determining their order of priority in a given situation (Gracia 2007, 7), and thus the deliberative method seeks to distance itself from an a priori rationality and creation of hierarchies that only make any exercise of prudence a futile one.

The deliberative method appeared after an initial phase expressed in terms of principles (Gracia 1991), now replaced by values (Gracia 2007, 2013). The method's earliest versions (Gracia 2001a, 2003, 2004) continued to use both languages: that of values, accompanied by facts, duties and norms, and that of principles, accompanied by consequences. Its simplified (Gracia and Rodríguez Sendín 2006) and fourth (Gracia 2011b) versions, however, only use the language of values. Furthermore, the latter version provides precise grounds for the method's meaning and characteristics, defines five steps in the argumentative process and includes a detailed theoretical reflection on deliberation presented in three orders: the biological or anthropological order (the human being as *animal deliberans*), the logical order (deliberation as a method for dialectical reasoning) and the ethical or moral order (deliberation in three internally related spheres: facts, values and duties) (Gracia 2011b).

The deliberative procedure as a tool for biomedical decision-making is thus structured as follows (Gracia 2011b, 125):

I. Deliberation on *facts*.

   1. Presentation of the case.
   2. Deliberation of the facts of the case:

      (a) What is the situation? (Diagnosis).
      (b) How is it going to evolve? (Prognosis).
      (c) What can be done? (Treatment).

II. Deliberation on *values*.

   3. Identification of the ethical problems presented by the case.
   4. Choice of the moral problem to be discussed.
   5. Determination of the values in conflict.

III. Deliberation on *duties*.

   6. Identification of the extreme courses of action.
   7. Search for intermediate courses of action.
   8. Choice of the optimal course of action.

IV. Test of the *consistency* of the decision.

   9. Test of legality.
   10. Test of publicity.
   11. Test of time.

V. Making the *final* decision.

This schematic overview can be supplemented and clarified by the explanation given in the simplified version, which structures the deliberative method in four successive levels:

Facts. The process must be based on the clinical facts. A sound factual basis is essential, amongst other reasons because errors at this stage will be perpetuated throughout all the subsequent stages (Gracia and Rodríguez Sendín 2006, 4, 2012, 4).

Values. Values can be defined as all those things that are important to a human being and which must be respected. The purpose of deliberation is to identify the best possible course of action when faced with a moral conflict, which is necessarily a conflict of values: this occurs when we deal with two or more contending values, each of which must be respected, or when we find it impossible to take such values into consideration at one and the same time (Gracia and Rodríguez Sendín 2006, 4, 2012, 4–5).

Duties. The purpose of ethics is to tell us what we should do: implement positive values, or if this is not possible, do the least possible harm to them. The outcomes to conflicts of values are known as "courses of action", and should never be limited to two extremes in which only one of the conflicting values is implemented and the other is harmed. As the most prudent and wise solution lies in finding a middle course of action, it is necessary to identify as many of these as possible in order to be able to identify the optimal one, which will achieve the best harmony between the conflicting values (Gracia and Rodríguez Sendín 2006, 4–5).

Norms. Ethics pursues the optimal solution, not simply any solution that is not bad (Gracia and Rodríguez Sendín 2006, 6). However, even when the purpose of the deliberative method is the ethical analysis of the case, at the end of the process the decision must always be tested with the legal norms, so as to ensure that it does not overstep the limits prescribed by law (Gracia and Rodríguez Sendín 2012, 6).

7.3 The Deliberative Method: A Critical Assessment

The deliberative method is unable to provide a solution to all the healthcare problems, and no does it attempt to do so. One of the main reasons for this are the limitations imposed by the very design of the procedure itself, such as the requirement to analyse only one ethical problem at a time (Step 4), which means that analysis of a second or further ethical issue involves returning to Step 3, choosing a new problem and then deliberating on it independently (Step 4_2).

The method is also open to errors in the way it is used, such as the enforced search for extreme courses of action (Level III, Step 6) knowing full well that the subsequent choice will have to be made from amongst the intermediate alternatives (Step 7). Even worse are its spurious or hypocritical uses, which create a risk of substantive irrelevance, where any conclusion deriving from the method is acceptable, with no need for external control or justification; or that of moral comfort, when a decision is taken as being "ethical" and "right" simply because it is an outcome of the method. These cases reveal even greater risks, such as that of accepting a rhetorical or pragmatic justification instead of looking for an authentically dialectical justification as a result of deliberation, or accepting a purely internal, procedural or intra-systemic justification that does not question the assumed concepts, values, duties and norms.

Taking the latest version of the deliberative procedure (Gracia 2011b) as our point of reference, I shall now assess the significance of each level and step, introducing suggestions that strengthen its justification and implementation.

7.3.1 The Facts of the Case (Level I)

Two steps or two levels? The facts are the starting point of the deliberative procedure. It is important to distinguish between the two steps within this level (Level I), namely the presentation of the case (Step 1), which does not involve any deliberation, and the beginning of the deliberation proper, in which the facts of the case are clarified and defined (Step 2).

The impure nature of the facts: facts and values. Differentiating between the factual level (Level I) and the axiological and normative levels (Levels II, III, IV and V) helps us to organize the deliberative procedure and the decision-making process. This distinction is an analytical or operational one, but does not express the positivistic separation between facts and values, since all facts are value-laden (Gracia 2010, 2013).

It is impossible to avoid the normative influence when characterising the facts, which are conditioned by the values, duties and norms of the case. They are not a mere description, but rather a selection of ethically relevant features, and in this sense cannot be established in advance, but are instead shaped by deliberation. The method thus acquires a certain hermeneutical tone (Gracia 2011b, 121).

Facts first. For the deliberative procedure to function properly it is necessary to clarify matters of fact that have been resolved unsatisfactorily, either as the result of a lack of certain empirical knowledge at the time of determination and deliberation on facts (Level I), or of confusing empirical matters with axiological or normative ones when identifying the ethical problems (Level II, Step 3), or even of having paid insufficient attention to certain facts that subsequently acquire greater relevance in accordance with the ethical issue that has been chosen (Level II, Step 4).

After Level 1 the uncontroversial facts cease to be the subject of deliberation and become premises in the axiological and normative deliberation (Levels II and III, respectively) (Perelman and Olbrechts-Tyteca 1969, 68–69). However, if a matter of fact has been unsatisfactorily resolved or new matters of fact arise in the subsequent levels we will have to return to the level of facts (Level I). The discursive theories of legal argumentation have resolved this issue in the form of transition rules (Alexy 1989, 206), whilst hermeneutics contemplates such a possibility ab initio as a way of determining the facts by means of a permanent dialogue between factual and normative elements.

7.3.2 Identifying Ethical Issues (Step 3)

Confusing facts with values. The axiological level poses the problem of managing the language of values. Moreover, the identification of ethical issues combines the language of values with that of duties, creating further difficulties. It is important not to confuse technical or clinical issues (Step 2), which pertain to the factual dimension (Level I) – e.g. whether an intervention is clinically indicated–, with ethical issues (Step 3), which pertain to the axiological dimension (Level II) – e.g. questioning the rightness or ethical significance of a patient's refusal of a clinically indicated intervention.

Patient competence as a factual issue and ethical issue. Determining a patient's competence can create confusion between matters of fact and matters of value. Determining a patient's competence is a technical issue that defines the course and the individuals of the decision-making process. However, failure to assess a patient's competence is in itself an ethical issue, one that can even raise a further ethical problem if a patient's competence remains undefined as the result of a lack of professional knowledge or suitable instruments for its assessment. These circumstances should nevertheless not be confused with the determination of competence, which is a purely clinical or technical matter. Put another way, answering the question of a patient's competence should be seen as a clinical or factual issue, and only the absence of an assessment can be considered an ethical one.

It is advisable to settle the ethical issue of a lack of assessment of a patient's competence before deliberating the subsequent steps, and to do so as a technical or factual issue (Level I, Step 2) before going on to choose the ethical problem to be discussed (Step 4). The absence of an answer to this question would require the deliberative body to reconstruct a tree of courses of action on the basis of the outcome of the assessment of competence.

7.3.3 Choosing a New Ethical Problem (Step 4)

If, once the final decision has been made (Level V), a further ethical problem still needs to be resolved (Step 3), we will have to commence the deliberation process from the beginning again by choosing a new ethical problem (Step 4) and repeating all the subsequent steps.

There may be times when, depending on the relation between the first problem to be chosen and those chosen subsequently, deliberation in the latter case may lack the ethical purity of that in the former, since it is hard to ignore the argumentations and conclusions pertaining to the previous issue.

7.3.4 Identifying Values in Conflict (Step 5)

Confusing values and duties. The language of values is not an easy one to use (Gracia 2013, 223–245). Benefit should not be described as a value, since it is really the outcome of an action taken to fulfil a duty (Level III). Furthermore, both of the conflicting values are potentially beneficial for the patient, and it is precisely the uncertainty as to what is truly beneficial in the case in question that gives rise to the ethical conflict experienced by the healthcare professional.

Values, principles and duties. In line with the above, and eluding the vague and abstract language of principles, neither beneficence nor non-maleficence should be considered as values (Level II), since they are really actions (do good, refrain from doing harm) performed in order to fulfil the duty to implement the corresponding positive value (Level III).

7.3.5 Tests of Consistency (Level IV)

The justification for testing consistency. Tests of consistency are a way of externally justifying the deliberative method, thereby reinforcing the normativity of the decision that has been made from an ethical, legal and practical standpoint. Refining this justification and normativity and adding two further tests of consistency will make them even more robust.

The ideal dimension and the real dimension. One of the functions of the tests of consistency is to conclude the deliberative procedure with a decision that can be applied *hic et nunc* and provides an answer to the professional's query. This connection with the real world by no means rules out the possibility of ideal solutions with no immediate practical consequences; indeed, the tests of consistency facilitate and justify the coexistence of both dimensions. Nevertheless, we should not identify the ideal dimension with deliberation on values (Level II), nor identify the real dimension with deliberation on duties (Level III) (Gracia 2011a, 89), because the normative level of duties allows for the simultaneous presence of both dimensions, the real and the ideal.

For this reason, and in accordance with the nature of the deliberative procedure, if the optimal course of action (Step 8, Level III) fails to pass any of the tests of consistency (Level IV) we will need to return to Step 8 in order to choose a new course of action from the intermediate ones remaining (Step 7), this being the second-best ethical option or a sub-optimal course of action (Step 8_2).

7.3.5.1 Test of Legality

Is this a legal decision? (Gracia 2004, 27). Once the case has been analysed from a purely ethical standpoint and the deliberative procedure has concluded, we also need to see whether the decision "… is also viable from a legal standpoint, because it cannot be generally considered prudent to make decisions in breach of law" (Gracia 2011b, 149).

Justification of the test of legality and its position at the end of the procedure. The test of legality is justified insofar as law is deemed to be the normative system that establishes the common criteria for harmonious coexistence in a given society, and one that claims to be universal, comprehensive and supreme.

The reason for introducing this test at the end of the procedure is to prevent it from supplanting the ethical reflection and to ensure the purity of the ethical deliberation, without identifying legal norms as authoritative arguments or making any *juris et de jure* presumptions of correctness – since it is enshrined in law it is right – thus avoiding using the law as a substitute for the moral answer.

What is legal is not necessarily ethically right, and thus it makes good sense to include the legal dimension at the end of the procedure. This also reinforces the distinction between the ideal discourse, which determines the normative horizon or the ethically optimal answer in the ideal community of communication (Habermas 1984, 175–183) or ideal deliberative community (Gracia 2011a, 87), and the real discourse, which fits the actual conditions of the case.

The meaning and scope of the term "legality". Legality should be understood in its broadest sense, including both legislation and case law. Legality is equivalent to currently applicable legislation, and thus excludes laws that have been repealed or have not yet come into force, as well as legislation that cannot be applied for material or territorial reasons. However, in order to enrich the deliberative procedure and the grounds for the final decision, it is possible to use legislative examples that are not applicable to the case to show courses of action considered to be ethically recommendable in other circumstances (e.g. legislation in other countries that provides a service or authorises an intervention that are not recognised in the territory in which the query has been made). For its part, case law is also of normative value, normally in the form of a precedent that guides the interpretation of the law or resolves legally relevant situations with no legal answer, and should therefore be contextualised in time and linked to the legislation applying at the time of the decision in order to avoid any anachronism.

Finally, the legal norms should be seen and interpreted systematically, relating the immediately applicable norm to the whole set of related legal provisions.

Divergence between the ethical and legal answers. What should we do when a decision fails the test of legality, or in other words, when the ethically optimal course of action is illegal? Legality is not the same as legitimacy. As the deliberative method is a process of ethical assessment, the reasonable and coherent answer would be to say at Level V that there is an ethically optimal decision (Step 8: Level III) which is not a legal one at the time the decision is made (because it fails to meet the requirements of Step 9: Level IV).

Since the method is intended to be practical and to provide a solution to healthcare professionals' conflicts, we should return to the intermediate courses of action (Step 7: Level III) and find a new suboptimal course of action that we then carry forward to Step 8 (Step 8_2 or Step 8_L) and subject the decision to the scrutiny of the tests of consistency (Level IV), and in particular the test of legality (Step 9), until we have determined a course of action that is both ethically right and legal. This will enable the professional to implement the committee's proposal, and the latter to fulfil its advisory function.

Although the final decision will always be up to the professional who has made the query, it would be unreasonable just to invite him or her break the law by suggesting a course of action that requires him or her to disobey a legal norm. On the other hand, if we include two answers at Level V – the ethically optimal but illegal course of action (Step 8_1 or Step 8_{-L}) and the ethical suboptimal but legal one (Step 8_2 or Step 8_L), this will provide the professional with full and accurate information on the committee's deliberation and more elements for analysis.

Furthermore, a dual answer of this kind fulfils an additional function by bringing to light a situation that may indicate the need for legal reform, thus extending the beneficial effects of the committee's work. If the existence of a legal norm does not in itself imply its fairness or material correctness, the committee's deliberation would provide arguments to justify a refusal to abide by the law or suggest legal reform.

7.3.5.2 Test of Publicity

Would you be able to publicly defend your decision, and would you have sufficient arguments to do so? (Gracia 2004, 27, 2011b, 150).

Justification and formulation of the test of publicity. An appropriate formulation of the test of publicity could be the Kantian transcendental formula of public law: "All actions relating to the rights of other men are unjust, if their maxim is not consistent with publicity" (Kant 1928, 381). Kant considers this not merely as an ethical principle, but also as a legal one. The discourse ethics has subsequently understood it as a ground rule for practical argumentation related to the requirement for openness and sincerity: all rules or arguments must be capable of being shown universally and openly (Alexy 1989, 130–131).

The rhetorical nature and the limits of the test of publicity. The test of publicity refers to the *possibility* of knowing the arguments on which the decision is based, but it is in no way designed to guarantee the decision's correctness or fairness. The prevailing element is the decision's *acceptability*, and in this sense it reflects a rhetorical, rather than an argumentative or dialectical, rationality. It would be possible to deliberate with the sole intention of persuading or gaining the assent of the audience, adapting one's arguments to ensure its adherence (Perelman and Olbrechts-Tyteca 1969, 4, 14, 23–25). Publicity is first and foremost aimed at the widespread knowledge and efficacy of the decision that has been reached through the deliberative process, and can be measured by how much the healthcare professional agrees or is satisfied with it (Perelman and Olbrechts-Tyteca 1969, 45). It is no guarantee, however, of the rightness of the decision that has been reached, unless what is morally right is identified with persuasion, efficacy or majority.

Furthermore, nor is publicity a guarantee of universalization, since it is sufficient to obtain the agreement, or absence of disagreement, of the majority of the audience.

The test of publicity ratifies its dimension of efficacy by operating with a utilitarian criterion: the greatest good for the greatest number.

Knowledge and understanding: intelligibility. Publicity is a guarantee of knowledge of or access to the decision and the arguments on which it is based, but not of it being able to be understood. The use of highly technical or specialized arguments would be enough to ensure that the addressees would be unable to understand them. To the test of publicity we should therefore add the requirement of *intelligibility*, intended as a guarantee of understanding and a ratification of respect for the addressee and his or her status as a valid interlocutor. Intelligibility does not so much refer to the way in which arguments and the subsequent decision are presented as to their being understood, and not only guarantees their ethical justification, but also their accessibility. Put another way, publicity is not just transparency: it is also accessibility and intelligibility.

All the arguments: sincerity and saturation. The test of publicity should be seen as a test of sincerity, honesty or truthfulness. *Sincerity* refers to the speaker's coherence, but is in itself insufficient, since it does not require one to say everything one knows, only that what is being said is believed by the speaker: every speaker may only assert what he or she actually believes (Alexy 1989, 188). *Honesty*, on the other hand, requires those participating in the deliberative procedure not to refer to justification they know is not valid (Aarnio 1987, 197), but is no guarantee that all the arguments are made known, since it does not require all the reasons, or everything the speaker believes, to be made public.

It is therefore not enough for arguments to be made public: the requirement is for *all* the arguments on which the decision is based to be made public. One way of ensuring this is to introduce the requirement of *truthfulness*, which reinforces confidence in the deliberative process by combining two characteristics or virtues: accuracy and sincerity (Williams 2002, 11). An alternative way is to fulfil the requirement of *saturation* (Alexy 1989, 245), according to which an argument (optimal course of action: Step 8) is only complete and justified if it contains all its premises. Saturated premises or arguments can be both empirical (Level I) and normative, referring to values and duties (Levels II and III), and are always open to being discussed or questioned anew.

7.3.5.3 Test of Time

Would you take the same decision in a few more hours or days? (Gracia 2004, 27). "It's a question of knowing whether the decision has been made as the result of a rush of emotion" (Gracia 2011b, 150).

Verification and the rebus sic stantibus *clause*. The purpose of the test of time is not so much to ratify the decision after a certain amount of time has passed as to subsequently verify that the deliberative procedure and decision are not insufficiently justified due to an incomplete analysis of the facts, values and norms as a result of the circumstances (emotional, severe restrictive or emergency circumstances).

According to the *rebus sic stantibus* clause, whilst the circumstances remain the decision taken as a result of the deliberative procedure will continue to be the final decision.

The principle of inertia and justification of any change. The test of time is no obstacle to a decision possessing certain characteristics proper to the specific circumstances in which it was adopted, nor the persistence over time of the criterion that has been followed, thereby acting as a guarantee of certainty. One reason to justify its inclusion and explain its significance would be the principle of inertia, the basis of the stability of spiritual and social life. Until proven otherwise, we must suppose that the stance adopted will continue forward into the future. Thus, once accepted a decision can neither be rejected nor abandoned without sufficient motive (Perelman and Olbrechts-Tytteca 1969, 106). Only change requires justification (Alexy 1989, 195–197), and those who claim a decision to be erroneous should provide further reasons for abandoning it and propose a new answer.

Furthermore, from the normative standpoint the test of time relates to the doctrine of precedent and the principle of justice (Perelman and Olbrechts-Tytteca 1969, 218–219; Alexy 1989, 274–279).

7.3.5.4 A New Test of Consistency: Test of Universalizability

Justification and formulation of the test of universalizability. Equality of treatment, the foundation stone of formal justice, is of the utmost importance in ethics and law. The test of publicity is unable to provide a guarantee of this requirement; this can be done, however, by means of a new test, that of universalizability. This test furnishes us with a discursive or argumentative justification, as opposed to the rhetorical nature of the test of publicity, which offers insufficient guarantee in societies that admit an unfair or discriminatory treatment of their members that could be defended in public but which would fail the test of universalizability. This new test of consistency gives us something different and more demanding than publicity, which only requires a decision to be known (and accepted) by the majority, whilst the test of universalizability refers to the universe of addressees and also to justice and equality of treatment.

Kant provides us with a formulation of this test in the first expression of his categorical imperative: "Act only in accordance with that maxim whereby you can at the same time will that it become a universal law" (Kant 1911, 421, 6–7). Furthermore, "I ought never to conduct myself otherwise than so that I could also will that my maxim become a universal law" (Kant 1911, 402, 8–9).

The meaning of this test can be explained by the rule of generalization, common in analytical theories of legal argumentation, which states that one may not refer to a value judgement that one would not generalize to cover other similar cases; or a willingness for role exchange, according to which one should be able to accept the consequences of decisions affecting others where one was in the position of those persons (Alexy 1989, 203; Aarnio 1987, 198).

7 Bioethics, Deliberation and Argumentation   101

Universalizability, dignity and rights. Although the deliberative method and practical decision-making require an individual analysis of each case, all of them nevertheless share a common starting point, namely the inadmissibility of unjustified discrimination. Furthermore, their connection to precedent or the requirements of equality invokes an external ethical justification, complemented by Kant's second formulation of his categorical imperative: "Act so that you treat humanity, whether in your own person as in that of any other, always at the same time as an end and never merely as means" (Kant 1911, 429, 10–13), as well as by the third: "In the realm of ends everything has either a price or a dignity. What has a price can be replaced by something else which is equivalent; on the other hand, what is above all price, and therefore admits no equivalent, has a dignity" (Kant 1911, 434, 31–34), or even the invocation of material ethical and legal universals such as human rights (Seoane 2008).

7.3.5.5 A New Test of Consistency: Test of Feasibility

Justification and meaning of the test of feasibility. The deliberative method aims to provide an answer to a clinical question by formulating a course of action that can be implemented. It must be possible to implement or apply the course of action stated in the final decision (Level V): if not, the method's practical purpose would be thwarted. We therefore have to introduce an additional test of consistency: that of feasibility (Alexy 1989, 205).

This test presupposes an empirical knowledge of the facts and their context, confirming the relevance of the factual dimension (Level I). It may not be possible, however, to implement the ethically optimal course of action (Step 8) due to a lack of personal, technical, financial or other resources. In this event, it will be necessary to offer a second-best option that can be implemented, in keeping with the method's practical intentions and in order to provide a satisfactory answer to the healthcare professional's query (by returning to Step 7 and choosing one or more new intermediate courses of action before proceeding once again to Step 8), since the case in question could be solved by adopting a sub-optimal course of action (Step 8_2 or Step 8_F).

An optimal decision that could not be implemented would be admissible as a theoretical exercise if a committee were using the method to draft a general recommendation, but not in the case of a specific consultation. The possibility of theoretical learning is parasitical to the latter: it goes against the purpose of the deliberative method to conclude the procedure by proposing an ethically wise course of action (Step 8) of no practical consequence because it fails to provide a solution to the real problem (Step 8 should lead to Level V, that of the final decision, after passing the test of feasibility in Level IV). It must therefore be possible to ensure the practical feasibility of the decision that concludes the deliberate process. Non-feasible courses of action are admissible when identifying intermediate courses of action (Step 7, Level III), and also, but to a lesser extent, when initially formulating the optimal course of action (Step 8, Level III), but it would make no sense whatsoever to adopt one as a single final decision (Level V) after passing the filter of the tests of consistency, one of which should be the test of feasibility.

The deliberative committee should be aware of the impossibility of implementing the optimal course of action when the situation being dealt with is a general one, but it might well not be in the case of a specific situation affecting a healthcare professional that has not been brought to its attention. In the first of these cases it could decide ex officio to continue its deliberations in order to come up with a second-best but feasible course of action (Step 8_2 or Step 8_F), but in the latter it should perhaps accept the optimal proposal and, when its unfeasibility is demonstrated, the healthcare professional can bring the case before the committee again, explaining why the initial decision cannot be implemented, these reasons then being taken into account at the level of facts (Level 1) of the new deliberative procedure.

Proposing ethical or normative ideals and deliberating on issues of justice. The explicit inclusion of the test of feasibility emphasizes ethical analysis and the potential identification of the optimal course of action as a normative ideal, whether it matches reality or not, and also keeps the committee's educational function without losing its practical orientation. Limiting itself to identifying optimal courses of action regardless of their feasibility would refute its practical purpose; restricting the scope of its ethical deliberations to fit the real circumstances of the case would condition the purity of its ethical analysis, modify the method's structure and steps and suppress the ethical requirement to seek the optimal course of action, since it would be limited to reproducing reality, thereby eliminating the possibility of imagining better models. The distinction between the ideal dimension and the real one explains why it may be impossible to implement the optimal course of action here and/or now, without depriving it of its relevance as a normative horizon. It serves to identify possible defects in a system or to propose changes to legislation or healthcare policy. The opposite, however, does not hold true: if the real conditions of the query are known from the outset, the committee is under no obligation to add a reflection on the ideal solution.

The test of feasibility requires equity and the fair distribution of resources to be taken into account. This should be seen as an advantage, because the method only allows one problem to be dealt with at a time and questions of justice may have been left to one side; now they can be taken into consideration. The course of action chosen would be the one that is the fairest, most equitable or most efficient: this may even come into contradiction with that chosen as the optimal course of action (Step 8, Level III) if the latter is based on other values or principles, such as autonomy or beneficence, instead of taking justice as the point of reference (Step 4, Level II).

7.3.6 The Nature of the Final Decision (Level V)

A recommendation, not a command. The decision that constitutes the end point of the deliberative procedure is a recommendation, a proposal based on the best decision and the best possible course of action. It is thus neither a command nor an order that is binding on the healthcare professional. Nor should it be interpreted as an exclusionary reason for following the course of action that has been chosen, but as a valid *prima facie* reason for considering that such a course of action is the best possible one, and as such, should be implemented.

Seeing the decision as advice or a recommendation is in harmony with a distinctive feature of healthcare ethics committees, consultative bodies that raise a claim to authority (of a moral kind: *auctoritas*) as the reason for their decisions being accepted and implemented, but which have no power (*potestas*) to impose them. The committee has the duty to communicate the outcome of its deliberations and the course of action it proposes (Level V) in order to advise healthcare professionals in the making of a decision, but not to assume their responsibility, since it is they who have the final say as to the solution of their problem (Gracia and Rodríguez Sendín 2012, 2). The committee provides the best ethical answer to the query, but can neither replace nor take away a healthcare professional's decision-making responsibility.

Unanimity, consensus and dissenting opinion. Prudent and wise deliberation is concerned with things that may be different, with what is contingent. The committee's arguments may be convincing and reasonable, but never apodictic or absolute: another reason, argument, course of action and decision is always possible. Deliberation is not tantamount to consensus, but rather to a prudent and wise decision (Gracia 2011a, 93). Far from requiring a decision to be unanimous, its essence is that a reasonable and prudent decision should be reached after an exchange of reasons and arguments has taken place.

The possibility of more than one decision at Level V may be the consequence, first of all, of a matter of content. When an ethically optimal course of action (Step 8, Level III) fails to pass the tests of consistency (Step 9, Level IV), it is necessary to formulate a new and sub-optimal course of action (Step 8_2) that can pass the tests and be adopted as the final decision (Level V). As has already been pointed out, the deliberative nature of the method requires communication not only of the final decision that has passed all the tests of consistency (Step 8_2, Level III, and Level V) but also of the ethically optimal decision (Step 8_1, Level III) that has failed to do so, meaning that there will be two decisions at Level V.

The existence of more than one decision at Level V may also come about for a subjective reason. When a committee's members hold differing opinions it will be impossible to come to a unanimous final decision. In such circumstances the final decision (Level V) must be adopted by a majority vote, with one or more different courses of action being considered optimal by a minority or even a single member. These must be communicated, in the form of dissenting opinions, to the healthcare professional making the query, so as to fully and truthfully reflect the deliberative procedure.

Both situations can be seen as positive: unanimity reinforces the authority and persuasive power of the decision, whilst its absence can be taken as example of the fertility of genuine deliberation, of the dialectical wealth existing within the committee and of the latter's plural nature and maturity (if it were to be considered as a symptom of dogmatism or intransigence we would have to talk of a pseudo-deliberative procedure or a failed deliberation).

A plurality of decisions of this kind is a true reflection of the deliberative procedure and enriches the answer to the query, adding an educational dimension to the committee's intervention. The healthcare professional who has to make the decision will have more than one option at their disposal, even though the committee's answer establishes an order of priority by proposing one of the decisions as the final one, either because it enjoys the greatest majority or because it is the only one to pass all the tests of consistency, putting forward the others as less optimal alternatives.

7.4 The Deliberative Procedure: A Reformulation

The deliberative procedure has been refined and enhanced through each of its successive versions to date. In this spirit, I put forward a further revision which I consider to be a fuller, more coherent and more accurate version of the one currently in use, respecting the existing structure and formulations:

I. Presentation of the facts.

    1. Presentation of the case.

II. Deliberation on facts.

    2. Deliberation of the facts of the case.

III. Deliberation on values.

    3. Identification of the moral issues presented by the case.
    4. Choice of the moral problem to be discussed.
    5. Determination of the values in conflict.

IV. Deliberation of duties.

    6. Identification of the extreme courses of action.
    7. Search for intermediate courses of action.
    8. Choice of the optimal course of action.

V. Deliberation on the consistency of the decision.

    9. Tests of consistency.

        (a) Test of legality.
        (b) Test of publicity.
        (c) Test of time.
        (d) Test of universalizability.
        (e) Test of feasibility.

VI. Making the final decision.

    10. Final decision.

This reformulation of the deliberative method also combines levels and steps. It begins by restructuring the factual level, differentiating between the new first stage and level (Level I) of the presentation of the case (Step 1), which is descriptive in nature, from the deliberative stage itself (Step 2), now considered to be a level in its own right (Level II). Although both steps belong to the factual stage, only the second one requires deliberation. Dividing this stage into two levels lends the method's structure greater clarity, since it now commences and concludes with two non-deliberative levels (Levels I and VI), reserving the deliberative procedure proper for the intermediate levels (Levels II, III, IV and V).

The level containing the tests of consistency (Level V, previously Level IV) changes its name to include a reference to deliberation, because these tests involve a true deliberative procedure rather than being a purely formal check. The five tests of consistency are seen as sections of a single step, namely the transition from (or transformation of) what is ethically optimal to the final decision.

The last level, that of making the final decision (Level VI, previously Level V) incorporates a stand-alone step (Step 10. Final decision), allowing the deliberative procedure to conclude with a new numbered step. Level VI, like the new Level I, is not a deliberative one. The final or definitive decision is more of a corollary to the procedure, which leads to the said decision as the conclusion of a practical syllogism that does not involve any deliberation within the procedure. The revised procedure thus places the deliberation on the facts, values and duties between these two non-deliberative levels.

References

Aarnio, Aulis. 1987. *The rational as reasonable. A treatise on legal justification*. Dordrecht: Reidel.
Alexy, Robert. 1989. *A Theory of Legal Argumentation. The Theory of Rational Discourse as Theory of Legal Justification*. Trans. Ruth Adler and Neil MacCormick. Oxford: Oxford University Press.
Aristotle. 1999. *Nicomachean Ethics*. Trans. Terence Irwin. Indianapolis: Hackett.
Aristotle. 1926. *Art of Rhetoric*. Trans. John Henry Freese. Cambridge, MA: Harvard University Press.
Aubenque, Pierre. 1963. *La prudence chez Aristotes*. Paris: Presses Universitaires de France.
Beauchamp, Tom L., and James F. Childress. 2013. *Principles of biomedical ethics*, 7th ed. New York: Oxford University Press.
Gracia, Diego. 1991. *Procedimientos de decisión en ética clínica*. Madrid: Eudema.
Gracia, Diego. 2001a. La deliberación moral: el método de la ética clínica. *Medicina Clínica* 117: 18–23.
Gracia, Diego. 2001b. La deliberación moral. *Boletín de la Academia Chilena de Medicina* XXXVIII: 29–45.
Gracia, Diego. 2001c. Moral deliberation: The role of methodologies in clinical ethics. *Medicine, Health Care and Philosophy* 4: 223–232.

Gracia, Diego. 2003. Ethical case deliberation and decision making. *Medicine, Health Care and Philosophy* 6: 227–233.
Gracia, Diego. 2004. La deliberación moral: el método de la ética clínica. In *Ética en la práctica clínica*, ed. Diego Gracia and Javier Júdez, 21–32. Madrid: Triacastela.
Gracia, Diego. 2005. The foundations of medical ethics in the democratic evolution of modern society. In *Clinical bioethics. A search for foundations*, ed. Corrado Viafora, 33–40. Dordrecht: Springer.
Gracia, Diego. 2007. Prólogo a la segunda edición (2007). In *Procedimientos de decisión en ética clínica* (1991), 1–8. Madrid: Triacastela.
Gracia, Diego. 2010. Philosophy: Ancient and contemporary approaches. In *Methods in medical ethics*, 2nd ed, ed. Sugarman Jeremy and Daniel P. Sulmasy, 55–71. Washington, DC: Georgetown University Press.
Gracia, Diego. 2011a. Deliberation and consensus. In *The SAGE handbook of health care ethics: Core and emerging issues*, ed. Chadwick Ruth, Henk ten Have, and Eric M. Meslin, 84–94. Los Angeles/London/New Delhi/Singapore/Washington, DC: SAGE.
Gracia, Diego. 2011b. Teoría y práctica de la deliberación moral. In *Bioética: el estado de la cuestión*, ed. Lydia Feito, Diego Gracia, and Miguel Sánchez, 101–154. Madrid: Triacastela.
Gracia, Diego. 2013. *Valor y precio*. Madrid: Triacastela.
Gracia, Diego y Rodríguez Sendín, Juan José (dir.). 2006. *Guías de Ética en la práctica médica 2. Ética en cuidados paliativos*, Madrid: Fundación de Ciencias de la Salud.
Gracia, Diego y Rodríguez Sendín, Juan José (dir.). 2012. *Guías de Ética en la práctica médica 6. Retos éticos en Atención Primaria*, Madrid: Fundación de Ciencias de la Salud.
Habermas, Jürgen. 1984. Wahrheitstheorien. In *Vorstudien und Ergänzungen zur Theorie des kommunikativen Handelns*, 127–183. Frankfurt am Main: Suhrkamp.
Kant, Immanuel. 1911. Grundlegung zur Metaphysik der Sitten (1785). In *Kant's gesammelte Schriften*, vol. IV, ed. Königlich Preussischen Akademie der Wissenschaften, 385–463. Berlin: Walter de Gruyter.
Kant, Immanuel. 1928. Zum ewigen Frieden (1795). In *Kant's gesammelte Schriften*, vol. VIII, ed. Königlich Preussischen Akademie der Wissenschaften, 341–386. Berlin: Walter de Gruyter.
Perelman, Chaïm and L. Olbrechts-Tytteca. 1969. *The New Rethoric. A Treatise on Argumentation.* Trans. J. Wilkinson and P. Weaver. Notre Dame: University of Notre Dame Press.
Seoane, José Antonio. 2008. La relación clínica del siglo XXI: cuestiones médicas, éticas y jurídicas. *Derecho y Salud* 16(1): 1–28.
Williams, Bernard. 2002. *Truth & truthfulness. An essay in genealogy*. Princeton: Princeton University Press.

Chapter 8
The Principle of Proportionality, Rights Theory and the Double Effect Doctrine

Juan Cianciardo

8.1 Introduction

In the following pages I will address the substantiation of the proportionality principle. More specifically, I will first explain some of Robert Alexy's ideas about this issue. Secondly, I will offer a critique of those ideas, particularly the grounds proposed by this German professor for the third sub-principle or sub-judgment of the principle of proportionality (which has its origins in Alexy's definition of principles as "an optimization of legal possibilities" – and also of factual ones, although the latter are used by this German author as grounds for the sub-principles of suitability and necessity). Thirdly, I will attempt to demonstrate that the principle of proportionality has a number of points in common with the old "double effect" doctrine: if such a connection were to be made more explicit and the doctrine of double effect were to serve as a wider basis for the proportionality principle (but not in its entirety, since the spheres in which they each operate do not fully coincide) it would be possible to avoid some of the problems faced by the proportionality principle in fulfilling the ambitious aim that inspired its creation, namely to guarantee the absolute nature of rights.

I thank Professor José A. Seoane, whose valuable suggestions and remarks have helped to make this work possible, and Professors Caridad Velarde, Alejandro Vigo, Mariano Crespo, Pilar Zambrano and David Thunder for their assistance. I also thank Professors Pedro Rivas and Juan B. Etcheverry, Aitor Rodríguez Salaverría, Ana Carolina Maluf, Ernst Thera, Luciano Laise and Marina Dandois for their support and cooperation.

J. Cianciardo (✉)
Universidad de Navarra, Pamplona, Spain
e-mail: jcianciardo@unav.es

I shall only include a brief initial reference to the principle of proportionality itself, taking as read its widespread dissemination in constitutional case law over the last 20 years or so, as well as to the definition and operation of the three sub-principles, and their use, with certain idiosyncrasies, in bioethics (Beauchamp and Childress 1989, 53; for a critical position, consistent with the one explained herein, see García Llerena 2012, *passim*).

Let us therefore begin with the initial reference. The proportionality principle (hereinafter, PrP), with its three dimensions of suitability, necessity and proportionality in the narrower sense, is increasingly recognized and applied in those legal systems in which the Constitution and rights play a major role, usually referred to as "neo-constitutional" systems (Prieto Sanchís 2004; Hirschl 2004; Serna 2005; Comanducci 2003; Cruz 2006). Its roots date back a long time – particularly within common law systems – , but its exact formulation is relatively new, and is due, at least initially, to the work of the German Constitutional Court. Coming closer to the present day, we can go so far as to say that from the 1990s onwards the expansion of the frontiers of neo-constitutionalism and the progress made by the proportionality principle have gone hand in hand,[1] its presence in comparative constitutional law nowadays being apparent (Landau 2010, 363). All the above explains the significant increase in the volume of research dedicated to this topic, some of whose findings are in its favor (cf., in general, the works referred to in note 2) whilst others are highly critical (Urbina 2012, *passim*; Webber 2010, 179–202; Cianciardo 2009, ch. 3; Barak 2012, 481–490). This criticism has focused above all on the lack of precision and ambiguity of the third (and crucial) sub-principle, and on the difficulties in substantiating the PrP as a whole. This article is intended to be a contribution to the latter debate, taking as its starting point the explanation and critiques of the ideas submitted by one of the PrP's most widely recognized champions, the German professor Robert Alexy.

The following question lies at the heart of the problem: what reason (or set of reasons) justifies the application of the PrP in a given legal system? This theoretical question has obvious practical consequences, since if such grounds constituted its legislative recognition – positive law –, their non-existence would suffice to prevent the principle from being applied; even if its grounds were the Constitution itself, constitutional reform would be enough to ensure that the principle could not be applied.

This question can be answered on a number of levels. In the first place, there is a systematic perspective, which entails examining why, at the present time and in a given legal system, laws must respect the PrP. This first approach, although necessary, does not however answer all the concerns arising with regard to the grounds for the PrP. A second approach has therefore been tried, an "extra-systematic" one,

---

[1] Cf., on this issue, Schwarze 1992, 680–702; Emiliou 1996, *passim;* Akehurst 1992, 38–39; Boyron 1992, 237–264; Barnes 1994, 495–499; Bermann 1978, 415–432; Braibant 1974, 297–306; Auby 1979, 227–238; Linares 2010, *passim*; Gavara de Cara 1994, 293–326; Gündisch 1983, 97–108; Alexy 2002a, 66–68; Willoughby 1929, *passim*; Georgiadou 1995, 532–541, Jiménez Campo 1983, 72.

within which two different levels of analysis may in turn be distinguished. One of these derives from a morphological study of the structure of legal rules (the logical perspective); more specifically, as we shall see later, from a comparison of the morphology of fundamental rights rules and that of the majority of other rules that make up a legal system. The other, at a philosophical level, originates in the reply to a more general question, one which inquires into the reasons for a requirement that can be considered to be universal, or at least very widespread (the ontological perspective). In the first case, we can mention the justification of the PrP in contexts that recognize fundamental rights, but do not expressly formulate them in a Constitution. In the second, however, it is more a question of inquiring as to the reasons explaining and justifying the requirement of proportionality in heterogeneous legal systems: this goes beyond the purely descriptive level and firmly enters the normative perspective.

Within this framework, the specific objective of this chapter is to expound and test Robert Alexy's answer to the initial question (What are the grounds of the proportionality principle?), which without a shadow of a doubt constitutes one of the most original and profound contributions to present-day constitutional law. In his first work on this subject, *A Theory of Constitutional Rights*, the author proposed a logical substantiation (Alexy 2002a, 66–68, also 1998, 21–33). More recently (Alexy 2011, 11–29), after confirming this initial proposal in general terms, he put forward a series of considerations intended to transcend legal logic and occupy a place in the field of philosophy of law, in line with his concern regarding the extremely "single-pointed" nature of his initial theory (Alexy 2007, 37–40).

To conclude this introduction, we must return to the ideas expressed at the outset, although now focusing on the underlying hypotheses of this study. The first of these is that Alexy's proposal is unsatisfactory for the following reasons: (a) his definition of principles as optimization requirements and, consequently, his conflictivist vision of the dynamics of principles; (b) as a result of the above, his proposal that balancing be used as a criterion for solving conflicts between principles; (c) his differentiation between fundamental rights principles and fundamental rights; (d) his insufficient consideration of the ontological perspective (by reducing the ontological perspective to a normative one). The second hypothesis is that the deficiencies we have pointed out lead to an incomplete formulation of the PrP, and more particularly of the third sub-principle. We say incomplete because it leads to a biased description of that sub-principle, and because it unavoidably implies that the PrP ends up being incapable of fulfilling its purpose, which is to proscribe absolutely any violation of rights. The third and final hypothesis is that it is possible to substantiate the PrP on better grounds than those put forward by Alexy, simply by adopting some aspects of his theory and omitting or amending others. This hypothesis is based on the following ideas: (a) the definition of principles as harmonization requirements; (b) the definition of balancing as a determination or specification of the content of each individual right; (c) the distinction between moral principles, fundamental rights principles, rules and constitutional rights; (d) the connection between the grounds of the PrP and those of the principle of reasonableness. The two latter hypotheses are nourished by the link between the PrP and the doctrine of double effect, which will be intuitively formulated here. The purpose of the proportionality principle coincides with that of the double effect principle, namely the legal-moral requirement in the case of the former and the moral requirement in that of the latter to acknowledge the absolute nature of human rights and moral absolutes, respectively. If this is the case, the sense of the principle as a whole lies in its capacity to avoid or transcend a mere cost-benefit balance.

Lastly, it is worth noticing that the reply to the key question of this work directly involves several of the main aspects of a constitutional rights theory that seeks to be comprehensive. It will therefore be unavoidably necessary to touch on such aspects during its development, although only to the extent necessary for this study.

8.2 An Approach to Substantiating the Principle of Proportionality: From Logic to Ontology

As Alexy explained, when answering the question put forward in the previous section, there are two opposing basic positions: "the thesis affirming that there is some kind of necessary connection between constitutional rights and the analysis of proportionality, and the thesis that holds, on the other hand, the non-existence of such a connection between constitutional rights and the principle of proportionality" (Alexy 2011, 11–12). The German professor defends the first thesis, which he calls "of necessity", and rebuts the second theory, which he calls "of contingency". His defense of the former is based on two explicit arguments: (a) that of the connection between the theory of principles and the PrP, and (b) that of the connection between constitutional rights and human rights (through principles), which leads us to uphold the "dual" nature of the former. Furthermore, there are two implicit grounds without which Alexy's thesis would not work: (a) the definition of constitutional rights principles as "optimization requirements"; and (b) the definition of constitutional rights as constitutional principles.

8.2.1 The First Necessity Thesis: The Connection Between Principles Theory and the Proportionality Principle

This thesis has two starting points familiar to those who are cognizant of Alexy's legal theory: the distinction between principles and rules, on the one hand, and the characterization of principles as "optimization requirements of factual and legal possibilities", on the other (Alexy 2012, 283–297; more extensively Alexy 2002a, 44–110). If both of these are taken as valid (we will not go into the difficulties posed by the identification of principles as optimization requirements, an issue that has been dealt with in a number of studies, amongst them Aarnio 2000, 593–602) the recognition of principles in a specific legal system would imply the existence and application of the PrP.

The reasoning behind this last point is that principles imply optimization, and optimization implies the PrP, since the three sub-principles in which the principle is specified (suitability, necessity and proportionality in the narrower sense) "together express the idea of optimization" (Alexy 2011, 13). While "the principles of suitability and necessity refer to optimization relative to the factual possibilities" (Alexy 2011, 13), i.e., saving costs that can be avoided (Alexy 2011, 15), the PrP in the narrower sense "concerns optimization relative to the legal possibilities", and implies balancing (Alexy 2011, 13). Let us now examine this in greater detail:

(a) In the case of the suitability sub-principle: "if a means M, adopted in order to promote the principle P_1, is not suitable for this purpose, but obstructs the realization of P_2, then there are no costs either to P_1 or P_2 if M is omitted, but there are costs to P_2 if M is adopted. Thus, P_1 and P_2 taken together may be realized to a higher degree relative to what is factually possible if M is abandoned. P_1 and P_2, when taken together, proscribe the use of M" (Alexy 2011, 13).

(b) Regarding the sub-principle of necessity, Alexy explains that "of two means promoting P_1 that are, broadly speaking, equally suitable, the one that interferes less intensively with P_2 has to be chosen. If there exists a less intensively interfering and equally suitable means, one position can be improved at no costs to the other. Under this condition, P_1 and P_2, taken together, require that the less intensively interfering means be applied" (Alexy 2011, 14–15).

(c) Finally, in those cases where costs are inevitable and there is a conflict between principles, the solution can only be obtained by balancing, which therefore becomes necessary. The PrP in the narrower sense "expresses the sense of balancing with regard to the legal possibilities" (Alexy 2011, 15), and is "identical to the rule known as 'the Law of Balancing'" (Alexy 2011, 15; also Alexy 2003, 433–449, 2007, 9–27). Under this rule: "the greater the non-satisfaction or impact of one principle, the greater the importance of satisfying the other has to be" (Alexy 2011, 15). If our aim is an accurate and complete analysis of such a balancing, the law of balancing needs to be improved by adding the "Weight Formula": "the weight formula defines the weight of a principle P_i in a concrete case, that is, the concrete weight of P_i relative to a colliding principle P_j (W_{ij}), as the quotient of, first, the product of the intensity of the interference with P_i (I_i) and the abstract weight of P_i (W_i) and the degree of reliability of the empirical assumptions concerning what the measure in question means for the non-realization of P_i (R_1), and, second, the product of the corresponding values with respect to P_j, now related to the realization of P_j" (Alexy 2011, 16). Alexy proposes the use of a triadic (discrete) scale to measure the quantities required for such calculations, to give a numerical representation of values such as slight, moderate and serious (Alexy 2011, 16–17).

The foregoing explanation leads Alexy to allege a necessary connection between principles theory and the PrP, based on the fact that "according to principles theory, principles are optimization requirements" (Alexy 2011, 19). Thus, "the proportionality principle with its three sub-principles of suitability, necessity and proportionality in the narrower sense logically follows from the nature of principles as optimization requirements, and the nature of principles as optimization requirements logically follows from the principle of proportionality" (Alexy 2011, 19).

The above raises a further question, which has already been mentioned, regarding the possible connection between constitutional rights and the theory of principles (or, to put it another way, proportionality analysis). It may be that the connection between principles and the PrP is accepted, but not the connection between principles and constitutional rights. If this were so, the consequence would be the affirmation of the contingency or positive thesis, according to which the connection between constitutional rights and principles (and, as a result of the latter, the connection to the PrP) depends on its acknowledgment by constitutional legislators. Alexy dedicates what he called the "second necessity thesis" to this issue.

8.2.2 The Second Necessity Thesis: The Connection Between Constitutional Rights and the Proportionality Principle

The question is therefore whether it is possible to derive from the recognition of constitutional rights their connection with principles and, consequently, their connection with the proportionality principle (given that the argument put forward above has served to demonstrate the relationship between principles and the PrP). The answer to this question is linked with our understanding of constitutional rights. If they are only positive law, the PrP's existence would depend, firstly, on the creation of constitutional rights by the legislator and, secondly, on the legislator's assignment of some kind of link between constitutional rights and constitutional principles. It does not seem reasonable to consider constitutional rights only as positive law and, at the same time, to hold that they cannot be created if the PrP is not simultaneously created (i.e. the creation of constitutional rights necessarily implies the creation of the PrP), since this would necessarily involve, as our analysis of the sub-principles has shown, having to bestow on the concept of constitutional rights an ideal sense or dimension that goes further than that of mere positivity (and is, therefore, incompatible with the merely positive nature alleged from this perspective). Therefore, if constitutional rights are only positive law, the necessity thesis would be false, and its opposite thesis, the contingency thesis, would be true.

Alexy seeks to answer these questions through two parallel strategies. The first is to resort to what he calls "the dual nature of constitutional rights", whilst the second implies asserting the necessary connection between the requirement for correction, on the one hand, and constitutional rights and law, in general, on the other. Each of these arguments suffices, in his opinion, to justify the second necessity thesis (Alexy 2011, 27).

8 The Principle of Proportionality, Rights Theory and the Double Effect Doctrine

According to Alexy, constitutional rights have a dual nature: actual (or factual) and ideal. In the first place, they are "positive law, that is to say, positive law at the level of the constitution" (Alexy 2011, 24). Along with this, they have an ideal dimension: "constitutional rights are rights that have been recorded in a constitution with the intention of transforming human rights into positive law" (Alexy 2011, 24). This is an "objective" intention, since this "is a claim necessarily raised by those who set down a catalogue of constitutional rights" (Alexy 2011, 24).

Alexy recognizes five main properties in human rights: they are moral, universal, fundamental, abstract and take priority over all other norms (Alexy 2011, 24, with reference to Alexy 2006, 17). The moral and the abstract properties are the key to understanding the necessary connection between constitutional rights and the PrP. The moral character of the former in the first place derives from the fact that only through this moral character can rights be justified, and since their validity depends solely on their justifiability, they therefore exist. In other words, their justification is the possibility of their being justified, which is moral, precisely on the basis of discourse theory: "the *Leitmotiv* of this justification is that the practice of asserting, asking, and arguing presupposes freedom and equality" (Alexy 2011, 25). In the second place, human rights are abstract because they refer "*simpliciter* to objects like freedom and equality; life and property; and free speech and protection of personality". In other words, their purpose is broad and indeterminate, and therefore (and this is the relevant consequence for the matter in hand) "as abstract rights, human rights inevitably collide with other human rights and with collective goods like protection of the environment and public safety" (Alexy 2011, 25). As a result, they need to be balanced.

The fact that human rights require balancing implies that constitutional rights also require balancing as a consequence of their dual nature, real (or factual) and ideal. Alexy holds that the ideal nature of human rights does not vanish once they have been transformed into positive law; "Rather, human rights remain connected with constitutional rights as reasons for or against the content that has been established by positivization and as reasons required by the open texture of constitutional rights" (Alexy 2011, 25). So how does the connection between constitutional rights and the PrP come into being? It does so in two ways:

(a) Potential connection. In cases where the legislator establishes a rule by means of which it solves a conflict between principles, then the formal principle of authority of the constitution demands that the rule be respected. The rule, however, may be extremely vague, obscure or ambiguous. In that case "the substantive principles standing behind it immediately come back into play" (Alexy 2011, 26) (e.g. when the rule is incompatible with constitutional principles) and balancing (i.e. the application of the PrP) becomes necessary.
(b) Actual connection. This arises in "all those cases in which constitutional rights norms, as set down in the constitution, have to be interpreted directly as principles" (Alexy 2011, 27; with reference to Alexy 2002a, 80–86, 109–117).

The second way in which Alexy substantiates the connection between constitutional rights and the PrP (i.e. the second variant of the necessity thesis) is based on the claim to correctness (Alexy 2002b, 35–36). The German author holds that "the claim to correctness, necessarily connected with constitutional review, requires that the decision of the constitutional court be as rational as possible" (Alexy 2011, 27). And the only way to satisfy this requirement is, in his opinion, by balancing or applying the PrP (Alexy 2011, 28; Cianciardo 2009, 107–125).

8.3 A Critique of Alexy's Position

Alexy's position, however, stands in need of correction, because it is based on erroneous assumptions that substantially weaken his conclusions. I will therefore now examine those aspects of Alexy's proposal that I consider less than solid, distinguishing between those which need to be better defined and those which should be put to one side.

8.3.1 First Critique: The Identification Between Constitutional Principles and Constitutional Rights

The first point of discussion is the identification between constitutional principles and constitutional rights. The hypothesis I aim to uphold is that constitutional principles are different from constitutional rights in their core or more interesting sense. Principles are starting points for legal reasoning; rights are specifications or determinations of what is established in principles, which are in turn determinations of what is set forth in moral principles (for a wider development of these ideas see Cianciardo 2007, Chap. 3). Along these lines, it has been rightly explained that the relevant point for establishing the extent of rights is not so much the abstract powers of their potential holders as the extent to which they create permitted, necessary or forbidden behaviors (Orrego 2010, 311–342). The truly crucial point is to establish if respect for a right requires any specific behavior on the part of its recipient. Within this task, any identification between principles and rights leads to the loss of the latter's essential attribute, namely their absolute nature, their strong resistance to any attempt to violate or impair them.

This absolute nature can only be preserved and made sense of by correctly distinguishing between principles and rights, and through the search for the proper determination of the latter. Only in this way does a legal discourse about rights become possible: "most assertions of right made in political discourse need to be subjected to a rational process of specification, assessment, and qualification, in a way that rather belies the peremptory or conclusory sound of '…have a right to…'. […] [T]hese claims assert two-term relations between a (class of) person and a (class of) subject-matter (life, body, free speech, property or ownership of property…). Before such assertions can reasonably be accorded a real conclusory force, they must be translated into specific three-term relations" (Finnis 2011, 218).[2]

From this precise point of view we perceive that the interest of Alexy's attempt to substantiate the PrP on what he calls the abstract nature of rights is very relative. Such grounds would allow (if the definition of principles proposed by Alexy is assumed) a connection between the PrP and principles, but not with rights, or, at least, not with their core meaning (the interesting point being therefore to determine the present and enforceable legal duty, akin to what occurs in the moral environment; with reference to bioethical principlism, see García Llerena 2012, 19).

8.3.2 Second Critique: A Conflictivist Vision of Dynamics Between Principles

Secondly, doubts also arise from the definition of constitutional principles as "optimization requirements" upheld by Alexy ever since his initial approaches to this issue (Alexy 1998, 12, 2005, 65–67), this being an argument on which he bases the first part of his "necessity thesis" (or "necessary inclusion") of the PrP in the recognition of principles). In my opinion, principles should be defined as harmonization requirements, that is, starting points for legal reasoning that seek respect for the principle according to which moral choices must target integral human fulfillment. A small difference, at first sight, but one with enormous consequences for the interpretation, cataloguing and justification of rights.

From Alexy's position, on the other hand, the open texture of constitutional principles necessarily leads to conflicts between them, as a reflection of conflicts that would arise between human rights (the ideal element to which he connects principles); this is an absolute starting point, with no fixed course to give it sense and, therefore, inevitably conflictive. In my opinion, on the other hand, the starting point for the interpretation of constitutional principles is not the existence of a conflict (in the sphere provided by the points of intersection between the factual assumptions of the different principles), or of several conflicts, but the aspiration to harmony

---

[2] The "translation" referred to by this author implies, according to his own words, specification of (a) the identity of the duty-holder(s) who must respect or give effect to A's right; (b) the content of the duty, in terms of specific act-descriptions, including the times and other circumstances and conditions for the applicability of the duty; (c) the identity or class-description of A, the correlative claim-right-holder(s) (in a Hohfeldian sense of 'claim-right'); (d) the conditions under which a claim-right-holder loses the claim-right, including the conditions (if any) under which the holder can waive the relevant duties; (e) the claim-rights, powers, and liberties of the claim-right-holder in the event of non-performance of duty; and, above all, (f) the liberties of the right-holder, including a specification of the limits of those liberties, i.e. a specification of the right-holder's duties, especially of non-interference with the liberties of other holders of that right or of other recognized rights. Since (f) involves specifying the duties of right-holder A, it necessarily involves a specification of the claim- rights of B, and this specification in turn requires a complete specification of points (a) to (f) in respect, now, of B; which will require a similar specification in respect of B's duties of non- interference with C…" (Finnis 2011, 246–247).

between each individual principle and the others (coordination). This aspiration is based on the nature of good held by the purpose of each principle (a nature that only exists as such if we accept that they can be harmonized with the other goods and the common good), and is expressed in each constitutional principle being the result of a determination of moral principles requiring, in turn, additional determinations coming from rules and judgments (Zambrano 2012, 135–157).

8.3.3 Third Critique: The Absolute Nature of Rights

Thirdly, Alexy denies the absolute nature of rights: "the conviction that there must be rights which even in the most extreme circumstances are not outweighed (…) cannot be maintained as a matter of constitutional law" (Alexy 2002a, 196). In my opinion, on the other hand, if their absolute nature is not asserted rights lose much of their attraction (one of the reasons for this being that it becomes difficult to justify their ideal dimension), and, in this regard, a proper justification of the PrP requires a distinction to be made in their classification according to whether or not they refer to moral absolutes, since their absolute nature is expressed differently in each case. In the first of these rights are absolutely intangible, whilst in the second they are not *prima facie* intangible. In this second case, however, a right is intangible once it, or more specifically its core or essential content, has been determined, without which it would be completely unintelligible (e.g. cases of hunger, life imprisonment, etc.).

Using a rather different terminology, Javier Hervada proposes a classification of "natural rights" as either "original" or "subsequent" (Hervada 2008, 114–116). While the former "are inherent to all men at any stage of human history" (Hervada 2008, 114), the latter "arise from human nature with relation to situations created by men" (Hervada 2008, 114). He explains that, "for example, both the right to life and its derivative, the right to take medication in order to preserve it, are original rights; self-defense, on the other hand, is a subsequent expression, since peace among men derives from human nature, but not an unjustified attack on another's life, which is something that derives from human will: once we have an attack created by man, defense appears as a subsequent expression of the right to life" (Hervada 2008, 114). Original rights are in turn divided into "primary" and "derived". The former are those rights "representing fundamental goods of human nature and those corresponding to their basic tendencies" (Hervada 2008, 115); the latter are "expressions and derivations of a primary right" (Hervada 2008, 115). According to Hervada, the importance of this distinction lies "in the diverse influence historicity exercises on each kind. While primary rights are constant and permanent, derived rights are subject to greater variability as regards their extent, because they are more dependent on historical conditions. Thus, the right to food and the right to education are more widespread in more developed environments that in less developed ones" (Hervada 2008, 115–116). Taking into account this second classification, we can better understand the assertions about the absolute nature of rights, *simpliciter* in the case of primary rights and *prima facie* in that of derived rights.

8.3.4 Fourth Critique: A Denial of the Incommensurable Nature of Rights

I also consider balancing as an appropriate technique for construing constitutional rights to be debatable. The human goods to which rights refer are totally or partially incommensurable, and therefore: (a) they cannot be balanced, either in whole or in part, and (b) any attempt at balancing leads to a manipulation of one of the rights and, therefore, to a denial of their totally or partially absolute nature.

The balancing test derives from a conflict between rights or between rights and goods, on the one hand, and the possibility of measuring them, on the other (Serna and Toller 2000, 10 ff.; Rivas 1999, 105–119). From this point of view, it is asserted that (a) all rights and goods are equal and equivalent to each other; and (b) "a necessary and casuistic balancing is [therefore] needed" (STC 104/1986, LG 5). Following this line of argument, it has been held that "[i]t is not a question of determining which is the more important good, since, exceptions apart, they are all equally so, particularly when the conflict arises between constitutional rights, but of deciding which of the two norms is more necessary, relevant or justified in order to protect the corresponding good or right" (Gascón Abellán 1990, 286; see also Prieto Sanchís 1990, 147–148).

The first problem of balancing is that it does not seem to be a rationally controllable activity, in any of its formulations. Nothing is said about the criteria that make it possible to decide in favor of one right or another. There is no guideline indicating why one of the two opposing freedoms (or, as the case may be, public goods) should prevail. Saying that the applicable norms have to be balanced is simply not enough, particularly when we consider the fundamental nature of human rights. On the other hand, and in connection with what has been said in the preceding section, balancing implies that one of the constitutional rights in conflict must be put to one side. As we have already explained, one of the main features of the discourse on human rights since their very inception is, precisely, their resistance to being set aside. Furthermore, it is a requirement of present-day constitutional doctrine: "the direct normative nature they today enjoy in every country with a Western legal culture makes it necessary to interpret them systematically, making their content internally compatible and interpreting each provision in accordance with the others" (Serna and Toller 2000, 32).

8.3.5 Substantiation of the Principle of Proportionality and the Doctrine of Double Effect

Here, I will first uphold the existence of a relation between the doctrine of double effect, principle theory and the substantiation of the PrP before going on to ask what the double effect is, how it is related to principle theory and, finally, how they are both related to the PrP.

8.3.5.1 The "Doctrine of Double Effect"

The so called "doctrine of double effect" or "of indirect agency" arises in the following context. All human acts provoke several effects, and it is inevitable that at least one of them is bad. Eating, for example, normally implies the death of at least one living creature. There are no human acts that do not have bad effects, although not all bad effects make the moral action producing them unlawful. It is not unlawful to produce bad effects entailing damages to property or, as in the above example, to sub-human realities. In these cases, it is enough that the good effect intended by the action be proportionate (there has to be a good reason for it) to justify the bad effect and therefore make the action a lawful one. Up to this point the main moral theories are in agreement. But henceforth two main avenues open up: (a) for one group of moral theories, proportionality (in the aforementioned sense, namely the existence of a reason with sufficient weight to displace the adverse weight of the bad effect) is the only criterion by which to judge the morality of an act, and, therefore, no bad effects are absolutely proscribed (i.e. absolutely proscribed to be used either as means or as ends), this group of moral theories usually being referred to as "proportionalist" due to their agreement on this point; (b) for the other set of moral theories, rooted in the tradition of classical thought (Boyle 1980, 527–538), on the other hand, there are bad effects that may never be legitimately pursued, or, in other words, there are certain bad effects whose intention (as means or ends) determines the illegitimacy of the action, regardless of how important their good effects are, these being known as "moral absolutes".

According to the latter position, however, such moral absolutes are not violated when they are not intended by the moral agent (Masek 2010, 567–585). In other words, when this type of bad effect is not desired by the agent, the act from which it derives is not necessarily (and provided certain conditions are met) unlawful. The doctrine of double effect seeks to distinguish, within this context, exactly what these conditions are.

The requirements are two, briefly. In the first place, the moral agent should not intend the bad effect but merely tolerate it (the bad effect is a collateral effect that is not directly desired; it is a case of "indirect" agency). Secondly, the directly intended good effect must be proportionate to the bad effect that is tolerated. If both conditions are fulfilled, the occurrence of the bad effect (or collateral effect) would not violate moral absolutes and would therefore not be unlawful.

A good example of the application of the doctrine of double effect to the field of bioethics is the case of "Vacco v. Quill", decided by the Supreme Court of the United States on 8 January 1997 (521 U.S. 793 (1997); Zambrano 2005, 200–211; McIntyre 2014). A group of doctors who regularly attended terminal patients and a number of patients themselves claimed that a New York rule criminalizing assisted suicide was unconstitutional. According to the petitioners, the law violated the right to equality protected by the Fourteenth Amendment of the Constitution. They argued that while patients undergoing critical medical treatments could dispose of their own life by merely rejecting or withdrawing such treatments, patients who were not undergoing vital treatments could only dispose of their lives through active assisted suicide measures, forbidden by law.

The Court rejected the petition of unconstitutionality (thereby reversing the positive decision of the previous instance). Justice Rehnquist, who led the majority opinion, constructed the holding on the basis of the doctrine of double effect. He argued in particular that the law did not violate the right to equality because the cases compared were essentially different from both an intentional and a causal perspective. From the intentional point of view, the Court argued that while physicians withdrawing vital care do not actively seek death, but rather only to respect the patients' constitutional right to reject medical care, active assisted suicide measures are undoubtedly intended to kill. On the other hand, from a causal point of view, while in the first case the cause of death is not the withdrawal of treatment but the underlying disease, in the second case the cause of death is the intake of lethal drugs.

8.3.5.2 The Doctrine of Double Effect and Constitutional Principles

Prof. Alejandro Miranda has suggested a possible link between constitutional principles and the doctrine of double effect: "in modern constitutions and declarations of rights, the provisions setting forth inviolable human rights are formulated in a similar way to the absolute moral prohibitions of the classical tradition. This has led, in more recent times, to a significant use of the doctrine of double effect in the constitutional protection of constitutional rights. Its application is an extensive one, in which the principle is used as a criterion to judge whether laws or other inferior rules conform to constitutional provisions".

This "extensive application" occurs because "it may be the case that a law, when aiming to achieve a legitimate good, nevertheless affects the exercise of practices that *prima facie* may be considered protected by a protected right, giving rise to a situation of double effect. In this context, judges have resorted in their reasoning to the same categories included in the traditional principle: distinguishing between the intention of the law and the collateral effects that it may provoke is legally relevant. In other terms, the principle has in this case served to distinguish between laws that directly impair the right protected by the constitution and laws that may limit or restrict that right as a collateral effect of pursuing other legitimate purposes. Laws of the first type are always unconstitutional, while those of the second type are not unconstitutional when through them the legislator pursues a sufficiently important good" (Miranda 2008, 513).

8.3.5.3 The Doctrine of Double Effect, Principles and the Proportionality Principle

This is the point where the doctrine of double effect connects with the PrP, because assessing the constitutionality of the regulation (or limitation or ruling) of a right (with a *prima facie* unlawful or bad effect) in order to achieve a legitimate effect has been done through the agency of the PrP. In other words, assessing whether the regulation of a right is constitutional or not has been made to depend on the regulation in question being: (a) adequate to achieve a purpose whose pursuit is permitted by the Constitution and is socially relevant; (b) efficient, in that the least restrictive of the rights involved amongst all those that are equally efficient to achieve the purpose; and (c) proportionate in the narrower sense, i.e. there is a proportionate balance between costs (regulation of the right) and the end pursued.

In my opinion, a thorough analysis of the above-mentioned connection would make it possible to provide elements for: (a) a better substantiation of the PrP; and (b) a more precise definition of the third sub-principle, particularly when the interpretation of absolute rights, i.e. the interpretation of rights recognized by principles that forbid a behavior regardless of the purpose of such a behavior or the circumstances surrounding it, is at stake.

Both advantages derive from the same argument. The purpose of the doctrine of double effect is to guarantee the protection of moral absolutes or, to put it another way, to ensure compliance with unexceptionable moral rules. The general theory of constitutional rights has a similar purpose: to guarantee the protection of human rights or, in other words, to ensure compliance with constitutional principles. From this we can draw two conclusions. The first is that this purpose operates, in both cases, as grounds for both doctrine and theory, on the one hand, and (correlatively) for the three sub-principles composing the doctrine as well as for the PrP on the other. The second conclusion is that the sub-principle of proportionality, as it is usually understood, does not allow the principle, as a whole, to fully comply with the purpose substantiating it, since there may be cases in which the balance between costs and benefits results in the justification of a cost that consists in the violation of a right. In an assumption of this kind, it is necessary to complete the assessment of the justification with a second assessment relating to the non-alteration of the right (i.e. to examine whether its essential contents have been violated or not). Only when this additional assessment has been introduced will it be possible for the principle to achieve the purpose for which it was designed (on this issue see Cianciardo 2010, 177–186).

8.4 The Need for an Ontological Justification: The Dynamics of Human Relations as the Foundations of Reasonableness and Proportionality

Taking all the above into account, two radical alternatives open up in respect of the necessity thesis. The first of these is that if proportionality is understood as balancing, as Alexy proposes, then the necessity thesis is fundamentally incorrect and even runs contrary to an effective enforcement of rights. On the other hand, if the sub-principle of proportionality in the narrower sense, as it is understood by Alexy (and most legal authors), is replaced by a qualitative assessment aimed at specifying the rights at stake in a conflict (in other words, one that establishes their reasonable scope of operation), the necessity thesis is basically correct. Put another way, we need to assess the extent to which the PrP is a requirement of constitutional principles and constitutional rights.

From our point of view, the recognition of constitutional rights involves the PrP for the following reasons:

1. The constitutional rights principles in which constitutional rights are recognized require determination. They require specification, as has been mentioned above, and are only *starting points* for legal reasoning.
2. The PrP allows us to assess the reasonableness of the legislator's determination (through rules established in laws) or that of a judge (through rules established in judgments) of a given constitutional principle.
3. The assessment of the reasonableness of a determination (of any legal determination, not only that created by the rules in respect of constitutional principles) is not only possible, but is also a requirement of law. In other words, law requires a distinction to be made between arbitrariness and determination. The latter links proportionality grounds with reasonableness grounds and also implies, although we are unable here to enlarge on this point, that: (a) the judge is able to link rules to constitutional principles and constitutional principles to moral principles; and (b) the sense of each rule is not definitively determined by its use (i.e. from the semantic point of view, the reference prevails over the meaning) (Zambrano 2012, *passim*).

A law will be consistent with the constitution or its interpretation will be constitutionally acceptable to the extent that it is reasonable. The assessment of reasonableness is, from this perspective, classificatory (Alexy 2002b, 26): unreasonable rules or unreasonable interpretations are not *legal* rules or interpretations. The problem of the substantiation of the reasonableness requirement may be redirected, from this perspective, to the issue of the substantiation of law. Determining why we demand that a law and its interpretation be reasonable is linked with the very idea of law in general, with the issue of its constitution and phenomenology and, beyond that, with an understanding of the legal phenomenon that provides an answer to this fundamental question: why law and not violence (Serna 2002, 321 ff.)? The anti-legalistic view, which is not limited to pointing out the insufficiency of the legal order for the effective fullness of human life, or criticizing the imperfection of any given legal instrument, but instead aims to completely annul any legal activity, is deeply contra-factual, on the one hand, and inhuman, on the other (for a punctilious critique see Cotta 1987b, p. 38). A satisfactory response should, on the contrary, be constructed on the basis of legal philosophy, a philosophy of the human being as a person. The reference to a person allows us to substantiate the ideal element of law and accounts for its formal structures, at least partially, and for law as a whole. From this standpoint, respect for and protection of the person become a structural element of the legal sphere, not merely the content or purpose of the norms (Serna 2002, 352–353). The legal avenue is by no means an irrelevant alternative for the subject: it is mandatory, since it is the only one that treats others as an end and not as a means because, in brief, it is the only one that respects their dignity. The expression 'dignity' refers to the eminence of human beings, their greatness, and its immediate translation has historically been a demand for unconditional respect as an end in itself, as Kant puts it. If respect is not unconditional, we are not talking about dignity, but about worth (which is relative). Dignity therefore implies "an absolute value, one that is not subject to any condition" (Serna 1998, 63–64).

The only *human* alternative is thus the legal one, which treats both the subject and others as an end, nor merely as a means, allowing an individual to convert his or her claim into a good that increases his or her human stature.

On the other hand, relations with others are only apparently conflictive. Indeed, while a confrontation with another person is also a confrontation with his or her claim, which prima facie may interfere with, be consistent with or be opposed to one's own claim, all circumstances considered, after the conflict of claims there will only be consistence or complementarity in an initially divergent relation (Serna and Toller 2000, 91–98 and *passim*). This is the reason why violence against another person, besides being a denial of our own personal nature, destroys the chance of transforming a claim considered to be true into an objective truth, appropriate to the subject him or herself, with frustrating consequences (Cotta 1987a, 47). The insufficient nature of the subjective feeling of truth leads to the acceptance of an objective and communicable truth.

This last issue is related to the material nature of the legal avenue. This avenue starts with the mutual recognition, not only of the dignified nature of the other, but also of the intangible nature of some goods, which deserve absolute respect. These intangible goods arising from recognition constitute, together with others created by agreement, the content of the legal alternative. Agreement has to be reached regarding everything else in which opposition exists. Each solution of opposing claims will thus entail, as well as recognition of the other party's equal capacity to make a claim, acceptance of the rational or cognizable nature of certain goods and the capacity to reach necessary agreements in respect of all remaining claims in which controversy or opposition exist.

The above provides the basis for the reasonable nature of each concrete solution for conflicts between opposing claims. They are reasonable solutions because they will be solutions to the extent that reasons can be put forward to uphold them, reasons that ultimately reside in certain intangible values or in agreements. These reasonable solutions do not only solve a one-off conflict, but are expressed in normative materials that serve as starting points for solving multiple substantially similar, i.e. analogous, conflicts between claims.

The legal channel imposes itself, therefore, by the very dynamics of human actions, in which human beings appear as an end in themselves, endowed with a rational nature. In the event of a conflict, the solution will be truly so if it is reasonable; thenceforth it must be considered a rule of behavior pursuant to which similar conflicts to the one solved will be settled. When this happens, the norm in question will have to be interpreted in one or another of its multiple reasonable senses. An unreasonable interpretation of a reasonable norm would entail a return to violence, and therefore constitute a violation of human dignity. Or, in other words, an unreasonable interpretation of the facts giving rise to claims or of the norms that act as starting points from which to solve the conflict inevitably leads to the frustration of the legal channel.

References

Aarnio, Aulius. 2000. Reglas y principios en el razonamiento jurídico. *Anuario da Facultade de Dereito de Universidade da Coruña* 4: 593–602.
Akehurst, Michael. 1992. The application of general principles of law by the Court of Justice of the European Communities. *British Year Book of International Law* 1981: 29–51.
Alexy, Robert. 1998. Derechos, razonamiento jurídico y discurso racional. In *Derecho y razón práctica*, 2nd ed, ed. W. Orozco, 21–33. México: Fontamara.
Alexy, Robert. 2003. On balancing and subsumption. *Ratio Iuris* 16: 433–449.
Alexy, Robert. 2006. Discourse theory and fundamental rights. In *Arguing fundamental rights*, ed. Agustín Menéndez and Erik Oddvar Eriksen, 15–29. Dordrecht: Springer.
Alexy, Robert. 2007. The weight formula. In *Frontiers of the economic analysis of law*, vol. 3, ed. Jerzy Stelmach, Bartosz Brożek, and Wojciech Załuski, 9–27. Cracow: Jagiellonian University Press.
Alexy, Robert. 2011. Los derechos fundamentales y el principio de proporcionalidad. *Revista Española de Derecho Constitucional* 31: 11–29.
Alexy, Robert. 2012. Rights and liberties as a concepts. In *The Oxford handbook of comparative constitutional law*, ed. Rosenfeld Michel and Sajó András, 283–297. Oxford: Oxford University Press.
Alexy, Robert. 2002a. *A Theory of Constitutional Rights*. Trans. J. Rivers. Oxford: Oxford University Press.
Alexy, Robert. 2002b. *The Argument from Injustice. A Reply to Legal Positivism*. Trans. S.L. Paulson and B.L. Paulson. Oxford: Oxford University Press.
Alexy, Robert. 2005. *La institucionalización de la justicia*, ed. J.A. Seoane. Trans. J.A. Seoane, E.R. Sodero and P. Rodríguez. Granada: Comares.
Auby, J.M. 1979. Le contrôle juridictionnel du degré de gravité d'une sanction disciplinaire. *Revue de Droit Public et de la Sciencie Politique en France et a l'étranger.* enero-febrero: 227–238.
Barak, Aharon. 2012. *Proportionality. Constitutional rights and their limitations*. Cambridge: Cambridge University Press.
Barnes, Javier. 1994. Introducción al principio de proporcionalidad en el Derecho comparado y comunitario. *Revista de la Administración Pública* 135: 495–499.
Beauchamp, Tom, and James Childress. 1989. *Principles of biomedical ethics*, 3rd ed. Oxford: Oxford University Press.
Bermann, George A. 1978. The principle of proportionality. *The American Journal of Comparatice Law* XXVI: 415–432.
Boyle, Joseph. 1980. Toward understanding the principle of double effect. *Ethics* 90–4: 527–538.
Boyron, Sophie. 1992. Proportionality in English Administrative Law: A faulty translation? *Oxford Journal of Legal Studies* 12: 237–264.
Braibant, Gregorie. 1974. *Le principe de proportionnalité*, Mélanges offerts a Marcel Waline. Le juge et le droit public, vol. t. II, 297–306. Paris: Librairie Générale de Droit et Jurisprudence.
Cianciardo, Juan. 2007. *El ejercicio regular de los derechos constitucionales. Análisis y crítica del conflictivismo*. Buenos Aires: Ad-hoc.
Cianciardo, Juan. 2009. *El principio de razonabilidad. Del debido proceso sustantivo al moderno principio de proporcionalidad*, 2nd ed. Ábaco: Buenos Aires.
Cianciardo, Juan. 2010. The principle of proportionality: The challenge of human rights. *Journal of Civil Law Studies* 3: 177–186.
Comanducci, Paolo. 2003. Formas de (neo)constitucionalismo: una análisis metateórico. In *Neoconstitucionalismo(s)*, ed. Miguel Carbonell, 75–98. Madrid: Trotta.
Cotta, Sergio. 1987a. *El derecho en la existencia humana*. Trans. I. Peidró Pastor. Pamplona: Eunsa.
Cotta, Sergio. 1987b. *Itinerarios humanos del Derecho*. Trans. J. Ballesteros. Pamplona: Eunsa.

Cruz, Luis M. 2006. *Estudios sobre el neoconstitucionalismo*. México: Porrúa.
Emiliou, Nicolas. 1996. *The principle of proportionality in European law. A comparative study*. London: Kluwer Law International.
Finnis, John. 2011. *Natural law and natural rights*, 2nd ed. Oxford: OUP.
Gavara de Cara, Juan Carlos. 1994. *Derechos fundamentales y desarrollo legislativo: la garantía del contenido esencial de los derechos fundamentales en la Ley fundamental de Bonn*. Madrid: Centro de Estudios Constitucionales.
García Llerena, Viviana. 2012. *El principialismo bioético y sus interlocutores. Notas críticas a la bioética contemporánea*. Granada: Comares.
Gascón Abellán, Marina. 1990. *Obediencia al Derecho y objeción de conciencia*. Madrid: Centro de Estudios Constitucionales.
Georgiadou, A.N. 1995. Le principe de la proportionnalité dans le cadre de la Jurisprudence de la Cour de Justice de la Communauté Européenne. *Archiv für Rechts- und Sozialphilosophie* 81: 532–541.
Gündisch, Herbert-Jürgen. 1983. Allgemeine Rechtsgrundsätze inder Rechtsprechung des Europäischen Gerichtshof. In *Das Wirtschaftsrecht des Gemeinsamen Marktes in der aktuellen Rechtsentwicklung*. Baden-Baden: 97.
Hervada, Javier. 2008. *Introducción crítica al derecho natural*. Buenos Aires: Ábaco.
Hirschl, Ran. 2004. *Towards juristocracy. The origins and consequences of the new constitutionalism*. Cambridge: Harvard University Press.
Jiménez Campo, Javier. 1983. La igualdad jurídica como límite al legislador. *Revista Española de Derecho Constitucional* 9: 71–114.
Landau, David. 2010. Political institutions and judicial role in comparative constitutional law. *Harvard International Law Journal* 51: 319–374.
Linares, Juan F. 2010. *Razonabilidad de las leyes. El "debido proceso" como garantía innominada en la Constitución Argentina*. Buenos Aires: Astrea.
Masek, Lawrence. 2010. Intentions, motives and the doctrine of double effect. *The Philosophical Quaterly* 60–240: 567–585.
McIntyre, Alison. 2014. Doctrine of double effect, The Stanford encyclopedia of Philosophy (Winter 2014 Edition), ed. E.N. Zalta, forthcoming URL = http://plato.stanford.edu/archives/win2014/entries/double-effect/
Miranda, Alejandro. 2008. El principio de doble efecto y su relevancia en el razonamiento jurídico. *Revista Chilena de Derecho* 35–3: 485–519.
Orrego, Cristóbal. 2010. Supuestos conflictos de derechos humanos y la especificación de la acción moral. *Revista Chilena de Derecho* 37–2: 311–342.
Prieto Sanchís, Luis. 1990. *Estudios sobre derechos fundamentales*. Madrid: Debate.
Prieto Sanchís, Luis. 2004. El constitucionalismo de los derechos. *Revista española de Derecho Constitucional* 71: 47–72.
Rivas, Pedro. 1999. Notas sobre las dificultades de la doctrina de la ponderación de bienes. *Persona y Derecho* 41–2: 105–119.
Schwarze, Jürgen. 1992. *European administrative law*. Luxembourg: Sweet and Maxwell.
Serna, Pedro. 1998. El derecho a la vida en el horizonte cultural europeo de fin de siglo. In *El derecho a la vida*, ed. Carlos Massini and Pedro Serna, 23–79. Pamplona: EUNSA.
Serna, Pedro. 2005. Presentación. In *La Constitución como orden de valores. Problemas jurídicos y políticos*, ed. Luis M. Cruz, XIII–XIX. Granada: Comares.
Serna, Pedro. 2002. *Proyecto docente y de investigación*, A Coruña.
Serna, Pedro, and Fernando Toller. 2000. *La interpretación constitucional de los derechos fundamentales. Una alternativa a los conflictos de derechos*. Buenos Aires: La Ley.
Urbina, Francisco. 2012. A critique of proportionality. *American Journal of Jurisprudence* 57: 49–80.

Webber, Grégoire C.N. 2010. Proportionality, balancing, and the cult of constitutional rights scholarship. *Canadian Journal of Law and Jurisprudence* XXIII–1: 179–202.
Willoughby, Westel Woodbury. 1929. *The constitutional law of the United States*. New York: Baker, Voorhis and Company.
Zambrano, Pilar. 2005. *La disponibilidad de la propia vida en el liberalismo político*. Buenos Aires: Ábaco.
Zambrano, Pilar. 2012. L'orizzonte comprensivo delle nostre pratiche costituzionali. *Ars Interpretandi* 12: 135–157.

Chapter 9
International Bioethics Committees: Conditions for a Good Deliberation

Vicente Bellver

One of the most relevant effects of bioethics emergence has been the spread of deliberative bodies on bioethics matters. Their features are very wide, depending upon the entity that creates them, their regional scope, the issues they deal with, the ruling strength of their agreements, etc. Among them, international bioethics committees are particularly relevant due to the huge impact of their work on global public opinion, as well as on the policies approved by governments all around the world. These bodies are presumed to adopt their decisions, as well as the other bioethics committees, after deliberating on facts and values (Gracia 2001).

I will begin this work by identifying the most relevant international bioethics committees working at present. In order to operate as deliberative bodies, they must fulfill two kinds of conditions: an adequate regulatory framework for their deliberation; and certain personal skills of the committee members. In this chapter, I will deal only with the first of these conditions, studying and comparing those regulatory aspects of the international bioethics committees that more directly affect their deliberation.

9.1 International Bioethics Committees

The concept "international bioethics committee" is vague. To begin with, any organization claiming to be a "bioethics committee" with authority beyond its borders is part of an "international bioethics committee". But, there are also other entities,

---

I want to thank Laurence Lwoff, Head of the Bioethics Unit Secretary of the Committee on Bioethics (DH-BIO), for the information provided about the DH-BIO.

V. Bellver (✉)
Department of Philosophy of Law, University of Valencia, Valencia, Spain
e-mail: Vicente.Bellver@uv.es

initiatives, practices, etc., which are not called "international bioethics committee", but contribute to the development of international bioethics.

In order to reduce its vagueness, I propose that the concept of "international bioethics committee" may be understood in a wide or a narrow sense. In a wide sense, an "international bioethics committee" is a (professional, academic, intergovernmental, non-governmental, etc.) body developing an activity (mainly the production of documents) that, in any sense, reflects the public opinion on bioethics matters and, at the same time, directly or indirectly contributes to form that public opinion on bioethics. In this wide sense, international bioethics committees are identified with bioethics organizations of global scope. According to Williams (2004, 32) "significant organizations in international bioethics include the following: United Nations and its agencies: WHO, UNESCO, UNAIDS, PAHO, etc., the Council for International Organizations of Medical Sciences (CIOMS), the World Medical Association, the Commonwealth Medical Association, the International Council of Nurses, the International Association for Law, Medicine and Science, the International Association of Bioethics and the International Conference on Harmonization of Technical Requirements for Registration of Pharmaceuticals for Human Use (ICH)".

In a narrow sense, international bioethics committees are bodies with ongoing deliberation activities, and international presence, mainly aiming to guide public policies on bioethics matters and citizen education on this subject-matter. Among international bioethics organizations, I propose to consider the following five organizations as international bioethics committees: the Committee on Bioethics (DH-BIO) of the Council of Europe; the UNESCO's International Bioethics Committee (IBC); the European Group on Ethics in Science and New Technologies; the Committee on Medical Ethics of the World Medical Association (WMA); and the Council for International Organizations of Medical Sciences (CIOMS) (Byk and Mémeteau 1996; Le Bris et al. 1997).

This selection may be subject to discussion. We may doubt that the CIOMS is a bioethics committee, since it lacks a stable structure and a body for bioethics deliberation. Nevertheless, I have chosen to include it because it has drafted a document which is a paramount reference for any regulation on research with human beings all over the world, and it is revised from time to time by working groups created for such purpose. There may also be doubts about excluding two bioethics entities with an international scope:

- The U.N. Inter-Agency Committee on Bioethics, created in 2003, aims at promoting the coordination of the works on bioethics performed by different United Nations' specialized agencies. Although it is called a bioethics committee, it may not be considered an international bioethics committee in a narrow sense, because its main role is coordination, not deliberation.
- The Global Health Ethics Unit at the World Health Organization (WHO) has three particularly relevant fields of activities. Firstly, it develops an intense work editing documents about bioethics matters of global interest, namely those related to the ethics of research and, particularly, to the ethical review of

health-related research. Specifically, in 2011, after a long consultation period, this body published the review of the Standards and Operational Guidance for Ethics Review of Health-Related Research with Human Participants (WHO 2011). Secondly, it is the permanent secretariat for the Global Summit of National Bioethics Advisory Bodies. And thirdly, it is the permanent secretariat for the Global Network of WHO Collaborating Centers for Bioethics (WHO 2010). Although its first responsibility may be deemed within an international bioethics committee scope, I consider its work is more inherent to a technical secretariat than a deliberative body.

In spite of including only five bodies in the category of international bioethics committee in a narrow sense, there are remarkable differences among them. Before beginning the comparison of their operation rules, it will be interesting to point out the characteristics of each of them.

9.1.1 *The Committee on Bioethics (DH-BIO) of the Council of Europe*

One of the main purposes of the Council of Europe is the protection of the human rights of the citizens of the Member States (art.1, b, Council of Europe 1949). In order to fulfill it, it has approved several legal instruments about human rights. The European Convention on Human Rights (1950) is the first and most important of the documents it approved, not only because it proclaims all civil and political rights, but also because it sets forth a supranational jurisdiction to which any citizen of a Member State may resort to file a complaint about a violation of the rights protected by the Convention.

Fully aware of the ambiguous effects of biomedicine on human rights, the Council of Europe decided to create a committee composed by delegations of each Member State to issue legal instruments in order to protect human rights against the risks posed by biomedicine. Although it has taken different legal forms (Ad hoc Committee, Steering Committee, and now, a subordinate body), this committee has worked since 1985, without interruptions. Its main achievement is the Convention for the Protection of Human Rights and Dignity of the HuCman Being with regard to the Application of Biology and Medicine: Convention on Human Rights and Biomedicine, also known as the Oviedo Convention (1997), and the four additional protocols to the Convention approved so far: on the Prohibition of Cloning Human Beings (1998); concerning Transplantation of Organs and Tissues of Human Origin (2002); concerning Biomedical Research (2005); and concerning Genetic Testing for Health Purposes (2008). Along with these legal instruments, which are binding for the ratifying States, recommendation drafts on bioethics matters were drawn, which have been subsequently adopted by the Committee of Ministers of the Council of Europe (which is the Council's decision-making body and is made up by the Ministers of Foreign Affairs of each Member State). It is worth highlighting that the Conventions and its Additional Protocols are "open for signature by the Member States of the Council of Europe, the non-member States which have participated in its elaboration and by the European Community" (Oviedo Convention 1997). Therefore, although its primary field of action is regional, it aims to be a model for other regions of the world (Roscam Abbing 1998).

This committee has the following main features:

- This is a stable and inter-governmental body, which is part of the Council of Europe's structure. Concretely, this is a subordinate body of the Steering Committee on Human Rights. It is permanently assisted by a technical secretariat, the Bioethics Unit, resorting under the Directorate General Human Rights and Rule of Law of the Council of Europe.
- Its most relevant mission is to prepare legal instruments intended for the protection of human rights facing biomedicine progress, which are adopted by the Committee of Ministers of the Council of Europe. But the terms of reference of the DH-BIO also include other responsibilities: (1) contributing to raising awareness and facilitating the implementation of the principles laid down in the adopted Oviedo Convention and its Additional Protocol, and (2) assessing ethical and legal challenges resulting from the development of the biomedical field (DH-BIO 2013). Therefore, it is focused on both, Law and ethics of the biomedical field.
- The texts drafted throughout these years and, particularly, the Oviedo Convention, made the Council of Europe become the main guarantor of human rights in the biomedicine field at international level (Bellver 2006). Those texts are based on the principle of human dignity, consistent with every document on human rights issued by the Council of Europe (Salako 2008). This foundation has been deemed useless (Macklin 2003) or ideological, by a bioethics sector in the academic area, while for another sector, this is the basis for an authentic global bioethics (Andorno 2009).
- It addresses all issues related to biomedicine, and is not restricted to a particular area. Its agenda is logically conditioned to the more worrying issues at the time being, and by the ability of reaching consensus (Romeo Casabona 2002). Precisely, some bioethics issues have never been addressed because they give rise to an -up to date- insuperable controversy (for example, abortion or euthanasia).

9.1.2 The International Committee on Bioethics (IBC)

It introduces itself as "the only global forum for reflection in bioethics". It is a stable body created within the UNESCO, integrated by 36 experts from all over the world. Although initially it dealt with genetic developments from an educative perspective (Kutukdjian 1994), soon it became interested in bioethics in general and completed the educative perspective with deliberation and consultation.

Among its most important documents, we may find three declarations part of the current international soft-law in bioethics: The Universal Declaration on the Human Genome and Human Rights (1997), the International Declaration on Human Genetic Data (2003), and the Universal Declaration on Bioethics and Human Rights (2005). These three declarations and, specifically, the most recent one, have been subject to strong criticism. It has been said that the UNESCO was dealing with issues corresponding to other agencies of the United Nations (particularly, the WHO); that UNESCO's declarations on bioethics are based on the concepts of human dignity and human rights, which are not globally admitted; and are defective in technical terms (Landman and Schüklenk 2005). These critiques have been thoroughly refuted (Andorno 2007; Yesley 2005).

The IBC is similar to the DH-BIO because both entities perform their work on bioethics from the legal perspective of the protection of human dignity and human rights. But it also shows very important differences from the latter:

- Its policies have universal scope.
- It may approve reports on any bioethics matter it deems pertinent. It is not necessary that a political body approves its documents, as in the case of the DH-BIO, with the exception of regulatory texts (declarations or conventions).
- The IBC is subject to certain control by an inter-governmental body, the Intergovernmental Bioethics Committee (IGBC), created by IBC Statute in 1998, and whose mission, without limitation, is to "examine the advice and recommendations of the IBC" (IBC 1998).

9.1.3 *The European Group on Ethics (EGE) in Science and New Technologies*

"The EGE is an independent and multidisciplinary consultative body to the European Commission, composed of up to 15 members" (Decision by the President of the European Commission 2011), which deals with ethics in science and new technologies. The EGE members serve in a personal capacity and have been appointed on the basis of their expertise and geographical distribution that reflects diversity in the European Union.

While the two former committees have drafted documents destined to become rules with international scope, the EGE restricts its functions to counseling the European Commission through the approval of opinions. Since its creation in 1997, it has approved 28 opinions. Almost every such opinion has addressed issues related to bioethics, although the EGE may address any ethical matter related to science and new technologies (Commission Decision 2010).

9.1.4 Committee on Medical Ethics of the World Medical Association (WMA)

While the three former international bioethics committees are part of intergovernmental organizations, the two remaining committees are associated to non-governmental organizations, of professional or scientific nature. Specifically, the WMA is an international organization founded in 1947, with the purpose of securing the highest ethical levels for the performance of medicine. This is composed by the national medical associations of 106 countries, which also support it on financial basis.

The WMA is globally known by four documents of ethical nature (Williams 2005). The first one is the Declaration of Geneva (1948), an update of the Hippocratic Oath, which is traditionally sworn or promised by the students at the time of their graduation and is bound together with the International Code of Medical Ethics, also approved by the WMA in 1949. The second one is the Declaration of Helsinki (1964), the best known document approved by the WMA, which sets forth the universal principles of ethics for biomedical research. These documents were the answer of the world medical profession to the atrocities done by doctors during the Second World War and the following years. These texts have been revised in several opportunities, but have maintained their brief and solemn nature.

Along with the two mentioned documents, we must also highlight two WWA declarations which are particularly significant: the Declaration of Tokyo: Guidelines for Physicians Concerning Torture and other Cruel, Inhuman or Degrading Treatment or Punishment in Relation to Detention and Imprisonment (adopted in 1975, and revised in 2005 and 2006) and the Declaration of Lisbon on the Rights of the Patient (adopted in 1981, amended in 1995 and revised in 2005).

In 1952, the WMA decided to create a stable Committee on Medical Ethics. At present, it is composed by 40 doctors from all over the world. This Committee, along with the WMA Council, has drafted declarations on medical ethics issues, which have been approved by the General Assembly. Since this committee only addresses issues related to medical ethics, this committee has not the same scope of the former, which addresses all bioethical issues and even ethical scientific and new technologies aspects, in general. On the other hand, although it is one of the counseling stable committees of the WMA, its official website does not show its regulation or its concrete activities. Thus, for example, it is not simple to find out the role performed by this committee in the subsequent amendments of the Declaration of Helsinki. There is no public information available about the role performed by the Committee on Medical Ethics, the WMA Council, some medical domestic associations and working groups created for certain works, in the drafting of declarations on medical ethics approved by the WMA Assembly. Since all these bodies participate in prior deliberation tasks, I understand that international bioethics committee considerations must be attributed to the WMA as a whole, and not to its Committee on Medical Ethics, in particular. This attribution is enhanced by considering that, in 2003, the Unit of Ethics of the WMA was created, which, in addition to encourage the activities related to ethics, carried out a revision of the WMA ethics-related declarations (Williams 2005).

9.1.5 The Council for International Organizations of Medical Sciences (CIOMS)

CIOMS is an international, non-governmental organization established jointly by WHO and UNESCO, in 1949. It represents the scientific community in the biomedical field and, at present, it is composed by 49 domestic and international organizations representing many of the biomedical disciplines, national science academies and the medical research councils. Its two large working areas are bioethics and health policies, and the development and use of medicines (particularly focusing on safety issues and adverse effects follow-up).

In the field of bioethics, CIOMS has approved two documents which are an international reference to regulate biomedical research, especially in developing countries, and are subject to revisions from time to time: the International Ethical Guidelines for Biomedical Research Involving Human Subjects (1982), whose last version was issued in 2002, and is being revised by a working group created for such purpose, at present; and the International Guidelines for Ethical Review of Epidemiological Studies (1991), which were revised in 2009. Also in the bioethics field, CIOMS approved the Principles of Medical Ethics Relevant to the Protection of Prisoners against Torture, which were adopted by the General Assembly of the United Nations in 1983.

CIOMS lacks a permanent body responsible for deliberation prior to the approval of any document related to bioethics. This work is usually commissioned to specific working groups, such as, the CIOMS Working Group on the Revision of the 2002 International Ethical Guidelines for Biomedical Research Involving Human Subjects, working at present.

9.1.6 International Bioethics Committees: Few and Heterogeneous

The revision of international bioethics committees allows us to reach two conclusions. First, while there are many inter-governmental, scientific, professional and academic organizations, and other organizations whatsoever that contribute to the development of international bioethics (which have been called international bioethics committees in a wide sense), only five of them may be considered international bioethics committees in a narrow sense because they maintain an ongoing deliberation about bioethics matters, aiming to contribute to citizen education and guiding public policies: they are the DH-BIO, IBC, EGE, WMA and CIOMS.

Second, there are big differences among those five organizations:

- Only one of them is an inter-governmental body as such: the DH-BIO. Other two are committees of experts created by an inter-governmental organization (UNESCO for the IBC, and the European Union for the EGE). The rest, WMA and CIOMS, are non-governmental bodies.
- Three of them are stable bodies for bioethical deliberation (the DH-BIO, IBC, EGE) and two are structures that fulfill several functions, among them, bioethical deliberation (the WMA and CIOMS).
- Two of them address all issues related to bioethics (the DH-BIO, IBC), other two only some bioethical issues (the WMA, issues related to medical ethics, although it tends to widen its scope, and CIOMS basically addresses research and health policies), and the EGE addresses issues beyond strictly bioethical matters, for example, ethical issues related to the development of science and new technologies.
- The DH-BIO can prepare legal instruments; the IBC can draft documents that may end up becoming international rules if they are adopted by the General Conference of the UNESCO; the WMA and CIOMS approve rules that, in spite of being professional guides, have had a decisive impact on the regulation of research all over the world; and the EGE approves reports which are not legally binding in any sense, whether directly or indirectly.

9.2 Regulation and Conditions for Good Deliberation of International Bioethics Committees

As I have just pointed out, there are important differences among the international bioethics committees. The most relevant could be the distinction between inter-governmental and non-governmental bioethics committees. The first ones make the rules creating these bodies available for the general public, as well as their operation rules. On the contrary, the two non-governmental organizations having bioethics committees share the same deficiencies: they do not have a stable body to coordinate and perform ethical deliberation before approving bioethics documents; and they do not make public the operation rules for the different bodies that participate in drafting bioethical documents, on general basis.

Below, I will discuss some regulatory aspects of the international bioethics committees that affect how the deliberation is carried out more directly.

9.2.1 Selection and Renewal of Members

- Committee of Bioethics (DH-BIO). Each Member State of the Council of Europe appoints a delegation to take part in the meetings of the DH-BIO. It is worth mentioning that the Council of Europe asks governments that the individuals appointed must have "appropriate expertise in the various aspects of bioethics and be able to consider these from a human rights perspective" (DH-BIO 2013). So, the human rights perspective of this body, which relates to ethics and Law, is reaffirmed.

This regulation has had several effects. In the first place, there are no limits to stay in the committee. Governments appoint and cease their representatives when and for as long as they deem convenient. Secondly, delegations may be integrated by only one or several experts. Thirdly, although committee members must be experts in bioethics, they are appointed on behalf of the States. Consequently, their professional skills are under the authority of the State they represent. A wide typology of delegations results from the combination of these three variables, including from numerous, stable and very active delegations in committee meetings, which defend the same bioethical criteria as the time passes by, to those single-member delegations, which frequently change their representative, has little participation, and change their bioethical criteria according to the current government they represent.

May a body whose members are chosen by the corresponding governments reach a good deliberation level? To begin with, the cooperative work of individual experts facilitates that the approved documents have a good level of legal, ethical and scientific coherence. They may also reach political consensus allowing the approval of legal instruments, as well as guidelines. But, may they carry out an authentic deliberation, which allows them to find out a fair decision for a certain biomedical area, although they are government representatives?

At the DH-BIO meetings, the delegations of non-Member States of the Council of Europe (Australia, Canada, Holy See, Israel, Japan, Mexico, and United States) participate in the discussions, without right to vote, as well as several intergovernmental organizations (for example, the European Union, WHO, UNESCO, etc.) or non-governmental organizations (for example, the European Science Foundation).

– International Committee of Bioethics (IBC). Its composition is completely different from the DH-BIO's. IBC members are not appointed by their governments, nor represent their governments, and their mandates are limited to two mandates of 4 years each. "The IBC shall be composed of 36 members appointed by the Director-General. Members shall be independent and shall act in their personal capacity. When making his choice, the Director-General shall take into account cultural diversity, balanced geographical representation and the need to ensure appropriate rotation. He shall also take into account the nominations for membership of the IBC received from the Member States of UNESCO, Associate Members and non-Member States which have set up a permanent observer mission to UNESCO" (art. 3.1, IBC Statutes 1998).

IBC members are appointed for 4-year mandates, and may be renewed only once. In any case, every 2 years half of the IBC members must be renewed (art. 6, IBC 1998). At the ICB meetings, the Member States or Associate States to the

UNESCO, as well as international organizations, may participate in the discussions, without right to vote.

The fact that appointments are personal facilitates the independent performance of their members. But, since the UNESCO Director-General must take into account the proposal received from the States, such independence may be limited.

- The European Group of Ethics (EGE) in Science and New Technologies. The President of the European Commission is responsible for appointing the 15 EGE members. They "are nominated *ad personam*. Members serve in a personal capacity and are asked to advise the Commission independently from any outside influence. The EGE shall be independent, pluralist and multidisciplinary. The identification and selection of EGE members will be made on the basis of an open call for expressions of interest. Additional applications received through other channels will also be taken into consideration in the selection procedure" (art. 3.2, EGE Mandate 2011).

The EGE system of appointments resembles the IBC's. In this case, members are elected for a 5-year period, and their term may be renewed for two additional periods. The fact that EGE members may stay in their positions for up to 15 years provides for the independence of their actions, but it may also make difficult the timely renewal of ideas and criteria within the EGE. It is up to the President of the Commission to find an adequate balance between continuity and renewal within the EGE.

Unlike the two above mentioned committees, the EGE is not open for the participation of observers, who can discuss issues, without vote.

- World Medical Association (WMA). As it has been said, one of the three permanent bodies of the WMA with counseling features is the Medical Ethics Committee, integrated by 40 doctors representing their corresponding domestic medical associations. There is no information available at the official WMA website about how the members of the Medical Ethics Committee are selected, or how its board operates, or its relations with the Ethics Unit of the WMA.
- Council for International Organizations of Medical Sciences (CIOMS). It does not have a stable body for deliberation on bioethical matters. Its work method in this area is based on the creation of ad hoc working groups to draft or revise any particular document. The official CIOMS website gives very little information about the CIOMS bodies involved in drafting documents on bioethics. Nevertheless, there is an exception, the Working Group on the Revision of the 2002 International Ethical Guidelines for Biomedical Research Involving Human Subjects, created by the CIOMS Executive Committee in 2011. At the heading of the 2002 International Ethical Guidelines, it is affirmed that they were "prepared by the Council for International Organizations of Medical Sciences (CIOMS) in collaboration with the World Health Organization (WHO)". At present, the WHO has clear guidelines that shape the process for drafting a guideline. Any document that contains a sentence stating that it was written "in collaboration with WHO" needs the upfront approval of the guideline review committee.

One of their conditions is a description of the process for drafting the guideline. As a consequence of this demand, the CIOMS Executive Committee meeting held on November 20th, 2013, decided to publish the information related to the selection process of working group members and working procedures (CIOMS Executive Resolution 2013).

In this case, the Working Group consists of ten members, one chair (President of CIOMS), four advisers (from the WHO, UNESCO, Council on Health Research for Development – COHRED and WMA) and one professional secretary. Members have one (or more) of the following backgrounds: physician, clinical researcher, medical ethicist, international health law expert young scientist or physician. The composition of the Working Group tends to reflect different cultural perspectives and reach a gender balance. A person taking the patient perspective is also present.

Undoubtedly, publishing the criteria adopted to create the Working Group and establishing its working procedure remarkably contributes to the transparency of the revision process of the 2002 International Ethical Guidelines. The fact that persons from less developed countries (Brazil, India, Senegal, Burkina Faso) have been incorporated to the Working Group helps to take into account these countries' point of view in the revision process. Anyway, most members of the Working Group (in their capacity of members, advisers or observers) come from high income countries and are familiarized with the culture of international organizations dedicated to issues related to health and biomedical research. Even under these circumstances, it cannot be completely guaranteed that the point of view of the less developed countries populations are taken into sufficient consideration at the time of setting forth ethical criteria under which research must be performed. But, it is difficult to imagine other measures that can be adopted in order to increase this guarantee, mostly if it is taken into account that all proposals of the Working Group are submitted to the opinion of the stakeholders.

9.2.2 Deliberation, Agreements and Dissenting Votes

Although when we discuss the bioethical committees in general terms, they are defined as counseling bodies with no ruling capacity, I understand that this consideration is not applicable to four of the five committees that we include in the category of international bioethics committees in a narrow sense. The DH-BIO and the ICB draft legal instruments which, once approved by the inter-governmental bodies from which they depend (the Council of Europe and UNESCO, respectively), become international rules. The DH-BIO, specifically, drafted the Oviedo Convention, which is a binding rule for all the ratifying States (at present, 29 of the 47 Member States of the Council of Europe). The IBC has approved three international declarations which, although they are not binding upon the States, they are part of the soft-law and constitute a world reference regulating issues related to genetics, and bioethics in general (Romeo Casabona 2014).

The WMA and the CIOMS, on their part, have drafted two documents, which are self-regulations by doctors and scientists, and hold such a strong symbolic significance that it seems unconceivable that a State approves rules on research with human beings infringing these instruments. I am referring to the Declaration of Helsinki and the Ethical Guidelines for Biomedical Research Involving Human Subjects, respectively.

Precisely, due to the ruling strength, whether direct or indirect, of some documents drafted by the international bioethics committees (except the EGE), it is even more important to know their corresponding method to deliberate, adopt agreements and, if possible, express dissent.

Before discussing the features of each committee, it is worth mentioning the fundamental role performed by the technical secretariats of each of these bodies. In the case of the DH-BIO, it is the Department of Bioethics, resorting under the Directorate General Human Rights and Rule of Law of the Council of Europe. In the case of the IBC, this role is fulfilled by the Bioethics Team of the Division of Ethics and Global Change. The EGE has the technical support of the BEPA (Bureau of European Policy Advisors) of the European Commission. In the WMA, this work may be done by the Ethics Unit and, in the CIOMS, it is not clear. These technical secretariats perform an essential role in the development of the international bioethics committee agenda and meetings. In order to adequately comply with their mission, it is necessary that they have sufficient financial and human resources and, at the same time, limit their role to technical assistance. The present crisis has adversely affected some of these secretariats, with drastic decreases in their resources, impairing the operation of their deliberative boards.

– DH-BIO. Within the Council of Europe structure, inter-governmental committees are bodies set up by the Committee of Ministers for technical and counseling purposes (Council of Europe 1949). They are subject to a detailed regulation as regards typology, characteristics and operation. They may fall in two categories: committees reporting to the Committee of Ministers (steering committees or ad hoc committees) and subordinate bodies reporting to steering or ad hoc committees (Council of Europe 2011).

The history of the present DH-BIO begins in 1985 when the Committee of Ministers created the Ad Hoc Committee of Experts on Bioethics (CAHBI), which, in 1992, became the Steering Committee on Bioethics (CDBI). In 2012, the CDBI lost its condition of steering committee and became a subsidiary committee, but with a particular status. According to its terms of reference, the DH-BIO shall carry out the tasks assigned to the Steering Committee on Bioethics (CDBI) by the Oviedo Convention under the authority of the Committee of Ministers. The rest of its work for the protection of human rights in the field of biomedicine assigned to it by the Committee of Ministers will be conducted under the supervision of the Steering Committee for Human Rights (CDDH) (DH-BIO 2013).

The plenary of the DH-BIO holds meetings twice a year, and its mission is defined under the terms of reference set forth by the Committee of Ministers every 2 years. Among its missions, we may highlight those established by art. 32 of the

Oviedo Convention: the periodical reexamination of the Convention text, as well as its additional protocols and, where appropriate, their revision.

When the DH-BIO votes amendments to the Oviedo Convention or any of its additional protocols, or proposals of new additional protocols to the Convention, it works as a steering committee, according to art. 32 of the Convention, whose paragraph 6 sets forth: "The Committee shall submit the text adopted by a two-thirds majority of the votes cast to the Committee of Ministers for approval".

For the rest of its work, the DH-BIO voting procedure follows the general rule for subordinate committees, which establishes: "Except on procedural matters, other committees shall not take decisions by voting. They shall state their conclusions in the form of unanimous recommendations, or, if this proves impossible, they shall make a majority recommendation and indicate the dissenting opinions" (Council of Europe 2011). Consequently, the DH-BIO has to approve its decisions unanimously or by the majority vote indicating dissident opinions, excepting when it is dealing with matters related to the Oviedo Convention.

This voting rule of the DH-BIO as a subordinate body of the CDDH offers two advantages. Firstly, since reaching a unanimous vote is more difficult, a higher exercise of deliberation and negotiation is demanded. Secondly, when agreements by unanimous vote are set aside, the majority vote must be accompanied by the dissenting vote. Thus, the majority position may not suppress dissenting voices. The CDDH, by receiving the majority vote indicating dissident opinions, may not only approve previous agreements, but it shall decide if it assumes the majority position or asks to reformulate it taking into account the dissenting positions, before sending it to the decision-making body, the Committee of Ministers.

DH-BIO deliberations are enriched by the contributions of the participants, as well as the observers attending its meetings. However, meetings are not public. To make them public would contribute not only to educate citizens, but also would make meetings more transparent and closer. Although meetings are not public, there is an abridged report of each one of them, which facilitates to learn about the issues addressed and the participants.

- IBC. Regular meetings are annual. Observers may participate in these meetings, which are public (Rule 26, Rules of Procedure 2001). The IBC may establish subsidiary bodies to develop their work.

- "The advice and recommendations of the IBC shall be taken by consensus, promptly made public and widely disseminated. Any member of the IBC shall have the right to record a dissenting opinion". (art. 7, IBC Statutes 1998) The rules of procedure concretize this general criterion in two ways: "21.1 The Committee shall endeavor to arrive at its decisions by consensus. In the event of a vote being taken, decisions shall be taken by a simple majority of the members present and voting. Each member of the Committee shall have one vote. 21.2 In the event of advice or recommendations to the Director-General of UNESCO concerning possible amendments to the Declaration, for submission to the General Conference, the decisions shall be taken by a two-thirds majority of the members present and voting" (Rule 21, IBC Rules of Procedure 2001).

- EGE. Just like the former committees, the EGE intends to adopt their decisions by unanimity, but "where an opinion is not adopted unanimously, it shall include any dissenting point of view" (art. 4.6, Commission Decision 2010). This committee sets forth an original rule regarding dissident votes: "Any EGE member who wants to dissent should announce this, at the least, at the meeting preceding the final meeting, so that there is enough time to discuss this dissent and, if possible, reach a consensus – and, if a consensus cannot be reached, to introduce the text of the dissenting view" (EGE Rules of Procedure 2011). This process offers two advantages: it prevents an eventual majority from disregarding minority positions without discussing them; and it avoids that minority positions show in their dissident vote arguments and positions that have not been previously discussed by the plenary.
- Working meetings are bimonthly and private, but the representatives of organizations may be invited in order to exchange points of view (art. 4.3, Commission Decision 2010). At this point, the contrast between the full disclosure of the meetings of the Presidential Commission for the Study of Bioethical Issues of the United States and the private nature of EGE meetings draws our attention.
- WMA. There is no information available at the official WMA website about how the members of the Medical Ethics Committee are selected and its working procedures. The information related to the operation of the Ethics Unit is not available, too. Therefore, it may not be reached by any person interested in learning about work procedures eventually conducting to the approval by the General Assembly of the recommendations on bioethics matters.
- Concretely, the contrast between the information available at the CIOMS official website about how the revision of the 2002 International Ethical Guidelines for Biomedical Research Involving Human Subjects was being carried out and the lack of information by the WMA about the process for the revision of the Declaration of Helsinki, that concluded with the approval of a new version by the 64th General Assembly of the WMA in Fortaleza, Brazil, in October, 2013, is striking. Only thanks to some incidental references about the revision of the Declaration of Helsinki, the general public has been aware of the creation of a working group, which extensively consulted stakeholders and justified the proposed revisions (Millum et al. 2013) and expert conferences were held in the revision process (Parsa-Parsi et al. 2013). It has been stated that the new version of the Declaration of Helsinki is legitimate, because it has been adopted and approved by democratic and transparent means by the General Assembly of the WMA (Parsa-Parsi et al. 2013). Without denying the legitimacy of a document which is, without any doubt, the cornerstone of ethical rules governing research on human beings, and acknowledging that many interested agents took part in the last revision process, I understand that this process has not been as transparent and public as it would be desirable for the most influential instrument at global level in the field of ethics for the research with human beings.
- CIOMS. There is no body within the CIOMS specifically dedicated to draft documents more directly related to bioethics, or information available at the CIOMS official website about its working procedures in the field of bioethics.

Nevertheless, in occasion of the revision of the 2002 International Ethical Guidelines for Biomedical Research Involving Human Subjects, the CIOMS agreed to publish notes, not only about the creation of the Working Group in charge of the revision, but also about their working procedure. Although they do not specify the number of meetings to be held, they illustrate the deliberation process to be carried out, which may be articulated in four steps. In the first one, the WG identifies the guidelines to be revised. In the second one, which is the most important, the WG takes into consideration the evidence available in this field and deliberates about each proposed change until reaching a consistent and consensual text. If this cannot be achieved, the existing version is maintained. In the third one, the document drafted by the WG is submitted to the consideration of the different interested agents and the general public. In the fourth, the WG studies the suggestions received and decides whether to include them. This new drafted version is the one finally submitted for the approval of the Executive Committee of CIOMS (CIOMS Executive Resolution 2013).

This deliberation process draws our attention because when the WG may not agree on a consensual text for a guideline they wanted to revise, the former text is maintained and a new one approved by a majority is not accepted. This encourages the members of the WG to make their best efforts to find a formula better than the previous one accepted by everybody, although many of them may consider it is not the best one.

It would have been desirable to clarify if the CIOMS Executive Committee may exercise its capacity to modify any aspect of the draft submitted by the WG, if it deems necessary, or it will only be limited to approve the text submitted.

9.3 Conclusion

The existing international bioethics committees all over the world are governmental and non-governmental in nature. Their mission ranges between ethical education of citizens and legal regulations for the States. Its perspective also fluctuates between ethics and Law. The rules regulating these bodies work may make deliberation, which is a fundamental activity for their good operation, easier or more difficult. The independence of their members, the transparency of their bodies, the publicity of their deliberations, the existing rules to reach agreements and state the dissenting votes, or the support given by technical offices to the respective international bioethics committees are some variables that may condition their deliberation. There is scarce research about the impact of the regulating rules of the international bioethics committees on their deliberation work. Since the texts they approve transcend globally (Aparisi 2005), it seems convenient to enhance that research in order to determine the rules that may better guarantee good deliberation.

References

Andorno, Roberto. 2007. Global bioethics at UNESCO: In defense of the universal declaration on bioethics and human rights. *Journal of Medical Ethics* 33(3): 150–154.

Andorno, Roberto. 2009. Human dignity and human rights as a common ground for a global bioethics. *Journal of Philosophy and Medicine* 34(3): 223–240. doi:10.1093/jmp/jhp023.

Aparisi, Angela. 2005. The globalization of bioethics: The tasks of international commissions. *Georgetown Journal of International Law* 37(1): 141–151.

Bellver, Vicente. 2006. *Por una bioética razonable. Medios de comunicación, comités de ética y Derecho*. Comares: Granada.

Christian, Byk, and Gérard Mémeteau. 1996. *Le droit des comités d'éthique*. Paris: Lacassaigne-ESKA.

CIOMS Executive Resolution. 2013. http://www.cioms.ch/index.php/12-newsflash/232-cioms-working-group-of-the-revision-of-the-2002-cioms-ethical-guidelines-for-biomedical-research. Accessed on 18 May 2014

Commission Decision. 2010. Commission decision of December 23rd, 2009, on the renewal of the mandate of the European Group on Ethics in Science and New Technologies (2010/1/EU). Official Journal of the European Union, 5.1.2010, L 1/8–9.

Council of Europe. 1949. Statute of the council of Europe. http://conventions.coe.int/Treaty/en/Treaties/Html/001.htm. Accessed on 22 May 2014.

Council of Europe. 2011. Resolution CM/Res (2011) 24 on intergovernmental committees and subordinate bodies, their terms of reference and working methods (Adopted by the Committee of Ministers, on 9th November 2011, at the 1125th meeting of the Ministers' Deputies).

Decision of the President of the European Commission. 2011. Decision of the President of the European commission of 10th Jan 2011, for the appointment of the European Group on Ethics in Science and New Technologies members for their fourth mandate (2011/C 12/04). Official Journal of the European Union, 15.1.2011, C 12/9.

DH-BIO. 2013. Terms of reference, committee of bioethics (DH-BIO) of the Council of Europe. DH-BIO/INF (2013) 3.

EGE Rules of Procedure 2011. Rules of Procedure of the European Group on Ethics (EGE) 2011–2016. http://ec.europa.eu/archives/bepa/european-group-ethics/docs/rules_of_procedure_en.pdf. Accessed on 14 May 2014.

Gracia, Diego. 2001. La deliberación moral. *Boletín de la Academia Chilena de Medicina* XXXVIII: 29–45.

IBC. 1998. Statutes of the International Committee on Bioethics of the UNESCO. Adopted by the Executive Board at its 154th Session, on 7th May 1998 (154 EX/Dec. 8.4).

IBC. 2001. Rules of Procedure of the International Bioethics Committee of UNESCO (IBC). SHS-503/01/CIB-8/4. Paris, Nov 23rd, 2001. http://www.unesco.org/new/en/social-and-human-sciences/themes/bioethics/international-bioethics-committee/. Accessed on 15 May 2014.

Kutukdjian, G.B. 1994. UNESCO international bioethics committee. *The Hastings Center Report* 24(2): 3–4.

Landman, Willem A., and Udo Schuklenk. 2005. UNESCO 'declares' universals on bioethics and human rights – Many unexpected universal truths unearthed by UN body. *Developing World Bioethics* 5(2): iii–iv.

Le Bris, Sonia, Maria Bartha Knoppers, and Lori Luther. 1997. International bioethics, human genetics, and normativity. *Houston Law Review* 33: 1363–1396.

Macklin, Ruth. 2003. Dignity is a useless concept. *BMJ* 327: 1419–1420. doi:http://dx.doi.org/10.1136/bmj.327.7429.1419.

Millum, Joseph, David Wendler, and Ezekiel J. Emanuel. 2013. The 50th anniversary of the declaration of Helsinki. Progress, but many remaining challenges. *JAMA* 310(20): 2143–2144. doi:10.1001/jama.2013.281632.

Oviedo Convention. 1997. Convention for the protection of human rights and dignity of the human being with regard to the application of biology and medicine: Convention on human rights and biomedicine, Oviedo 1.4.1997, http://conventions.coe.int/Treaty/en/Treaties/Html/164.htm. Accessed on 15th May 2014.

Parsa-Parsi, Ramin, Jeff Blackmer, Hans-Jörg Ehni, Torunn Janbu, Otmar Kloiber, and Urban Wiesing. 2013. Reconsidering the declaration of Helsinki. *The Lancet* 382(9900): 1246–1247. doi:10.1016/S0140-6736(13)62094-2.

Romeo Casabona, C. 2002. El Convenio Europeo sobre Derechos Humanos y Biomedicina: sus características y sus repercusiones en el Derecho español. In *El Convenio Europeo sobre Derechos Humanos y Biomedicina. Su entrada en vigor en el ordenamiento jurídico español*, ed. C. Romeo Casabona, 1–18. Granada: Chair on Law and Human Genome Comares, Bilbao.

Romeo Casabona, C. 2014. La construcción del Derecho aplicable a la genética y a la biotecnología humana a lo largo de las dos últimas décadas. *Revista de Derecho y Genoma Humano*. Extraordinary edition: 27–52.

Roscam Abbing, H. 1998. The convention on human rights and biomedicine: An appraisal of the council of Europe convention. *The European Journal of Health Law* 5(4): 377–387.

Salako, Solomon E. 2008. Informed consent under the European convention on biomedicine and the UNESCO declaration on bioethics. *Medicine and Law* 30(1): 101–113.

WHO. 2010. Global network of WHO collaborating centers for bioethics. Terms of reference. Adopted on Wednesday 26 May 2010, http://www.who.int/ethics/partnerships/global_network/en/. Accessed on 1 June 2014.

WHO. 2011. Standards and operational guidance for ethics review of health-related research with human participants, http://www.who.int/ethics/publications/research_standards_9789241502948/en/. Accessed on 1 June 2014.

Williams, John R. 2004. The promise and limits of international bioethics: Lessons from the recent revision of the declaration of Helsinki. *Journal International de Bioéthique* 15: 31–42. doi:10.3917/jib.151.0031.

Williams, John R. 2005. The ethics activities of the world medical association. *Science and Engineering Ethics* 11: 7–12.

Yesley, M. 2005. What's in a name? *The Hastings Center Report* 35(2): 8.